T0322013

Understanding Weightless

Essential for getting to grips with the Weightless standard for M2M communications, this definitive guide describes and explains the new standard in an accessible manner. It helps you to understand the Weightless standard by revealing its background and rationale. Designed to make clear the context and the fundamental design decisions for Weightless, and to provide a readable overview of the standard, it details principal features and issues of the technology, the business case for deployment, network performance and some important applications. This informative book guides you through the key decisions and requirements involved in designing and deploying a Weightless network. Includes a chapter on applications, explaining the relevance of the standard and its potential. Written by one of the lead designers of Weightless, this is an ideal guide for everyone involved with the standard, from those designing equipment to those making use of the technology.

WILLIAM WEBB is one of the founding directors of Neul, one of the core design team involved in Weightless and the CEO of the Weightless SIG. Prior to this William was a Director at Ofcom. He is a Fellow of the Royal Academy of Engineering, the IEEE and the IET, where he is a Vice President, and the author of multiple books, papers and patents.

Understanding Weightless

William Webb
Neul, Cambridge, UK

CAMBRIDGE
UNIVERSITY PRESS

CAMBRIDGE
UNIVERSITY PRESS

Shaftesbury Road, Cambridge CB2 8EA, United Kingdom

One Liberty Plaza, 20th Floor, New York, NY 10006, USA

477 Williamstown Road, Port Melbourne, VIC 3207, Australia

314–321, 3rd Floor, Plot 3, Splendor Forum, Jasola District Centre, New Delhi – 110025, India

103 Penang Road, #05–06/07, Visioncrest Commercial, Singapore 238467

Cambridge University Press is part of Cambridge University Press & Assessment, a department of the University of Cambridge.

We share the University's mission to contribute to society through the pursuit of education, learning and research at the highest international levels of excellence.

www.cambridge.org
Information on this title: www.cambridge.org/9781107027077

First published 2012

A catalogue record for this publication is available from the British Library

ISBN   978-1-107-02707-7   Hardback

Contents

Preface

This book is a guide to the Weightless technology. The definitive documentation for Weightless is the Weightless standard as published by the Weightless SIG and this book in no way is intended as an alternative. However, standards documents are not designed for readability and it is often helpful to start with a more descriptive text. This book is designed to explain the context for Weightless, the key design decisions and to provide a readable overview to the standard. That may be sufficient for some, and for those that will go on to read the standard they will better understand the details that they find there.

This book was written to accompany version 0.6 of the standard. This is referred to throughout the book simply as the *standard*. Of course, this standard will evolve and it is envisaged that this book will be updated to accompany major revisions to the *standard*.

Why the name 'Weightless'? Like many standards such as Zigbee and Bluetooth, the name chosen is often more whimsy than descriptive. In this case, it came indirectly from making the point that this was not a new generation of cellular standard, measured in generations (2G, 3G . . .). Indeed, it was outside of this generation pattern and so perhaps zero-G. Zero-G is often associated with no gravity leading to weightlessness . . . At the time it was conceived most thought that a different name would emerge during the early standardisation process, but as is often the case names acquire a momentum all of their own, even those that have no weight.

Acknowledgements

Any standard is the work of many individuals and companies – Weightless is no exception. As always there are too many to mention them all here, but my thanks go especially to the 'core' design team who conceived the Weightless standard and laid down its key operational parameters early in 2011. These include Robert Young, Tim Newton, Alan Coombs, Carl Orsborn and Neil MacMullen. Especial recognition is due to James Collier for having the vision to see the potential and the leadership to make it happen and to Vanessa Price for making sure we had plenty of coffee and biscuits.

About the author

William is one of the founding directors of Neul, a company developing machine-to-machine technologies and networks, and the CEO of the Weightless SIG, the body developing the Weightless standard. He was one of the core design team involved in Weightless and instrumental in establishing the SIG.

Prior to this William was a Director at Ofcom where he managed a team providing technical advice and performing research across all areas of Ofcom's regulatory remit. He also led some of the major reviews conducted by Ofcom including the Spectrum Framework Review, the development of Spectrum Usage Rights and most recently cognitive or white space policy. Previously, William worked for a range of communications consultancies in the UK in the fields of hardware design, computer simulation, propagation modelling, spectrum management and strategy development. William also spent three years providing strategic management across Motorola's entire communications portfolio, based in Chicago.

William has published 11 books, 80 papers and 18 patents. He is a Visiting Professor at Surrey University and DeMontfort University, a member of Ofcom's Spectrum Advisory Board (OSAB) and a Fellow of the Royal Academy of Engineering, the IEEE and the IET where he is a Vice President. William has a first class honours degree in electronics, a PhD and an MBA. In his spare time he is a keen cyclist having ridden Land's End to John O'Groats, the route of the Tour de France and completed the gruelling Cent Cols Challenge, riding over 100 Alpine mountain passes in 10 days.

1 The world of machine communications

1.1 What defines a machine?

Wireless communications are widespread, but tend to be used predominantly by people. Mobile phones allow people to talk or send emails, Wi-Fi systems allow people to surf the Internet from a laptop and Bluetooth links allow people to use cordless headsets. There is an entirely different class of applications for devices that do not directly have users and whose communications are not instigated by people. A good example of this is a smart electricity meter. It might send meter readings to a database every hour. It has no direct linkage with any person – although indirectly it makes their life somewhat better by enabling smart grids and automating the meter-reading process.

There are so many different applications and machines that a clear definition of one is not possible. Broadly, these are devices where transmissions occur due to the function of the machine rather than any person. They send information not to another person but typically to a database within the network from where it can be processed by other machines. Of course, sooner or later someone benefits from the service provided, but typically not from the radio transmission itself. Applications include automotive engine management updates, healthcare monitoring, smart city sensors and actuators, smart grids, asset tracking, industrial automation, traffic control and much more.

This type of communications is often referred to as machine-to-machine (M2M). This can be confusing as it might be taken to imply one remote device talking to another remote device (e.g. one smart meter talking to another smart meter). In practice, the communications are typically machine to network, or machine to control node. In this book the terminology 'machine communications' or 'machine network' is used rather than M2M.

1.2 Range: short or long?

Machine communications fall into two key types as regards the use of wireless communications – short or long range. Short-range applications are typically those that occur within a building such as a home or office. This might include wireless control of a building's lighting from a central controller within the building. The range of such applications is typically around 100 m, although this can be extended using mesh-architectures or repeaters within the building. This type of solution is sometimes called a home area network (HAN). There is a range of technologies already available for HANs including Zigbee, Bluetooth, Wi-Fi and some proprietary solutions.

Long-range applications are those that need to communicate wherever they are and sometimes as they move around. Automotive is a good example of such an application. They often require near-ubiquitous coverage of a country or even globally. As a result, they require the deployment of a cellular-like network and many similar arrangements as for cellular communications including billing and roaming.

This begs the question as to why not use cellular networks for long-range machine communications. Cellular networks are already used for some machine applications such as monitoring vending machines and within some cars. They bring the benefit of widespread coverage, good availability and widely available components. But cellular is far from ideal for machine communications. Issues include:

- Coverage is not perfect, especially within buildings.
- Terminals cannot run off batteries for extended periods – more than a week is problematic.
- Cellular networks are not well adapted to short messages and so are very inefficient for most machine applications.
- Treating each terminal as a subscriber adds costs including SIM cards, expanded billing systems and more.
- Cellular networks are moving towards higher data rates and away from the functionality required for machine communications.

These issues mean that while cellular has addressed a small, high-value segment of the machine market it cannot be used for the much broader marketplace. Indeed, if it were well-suited to this it would have been adopted years ago.

Determining whether an application is short or long range is not always simple. For example, a smart meter in a home could communicate via the HAN and then the home broadband connection into the network. However, this is complex to set up and leaves the energy supply companies vulnerable to the home user changing their HAN, or even just the password on their Wi-Fi router, and disabling the smart meter. For that reason, long-range communications are preferred for many devices even within the home.

Weightless is a standard for long-range machine communications. Hence, it is expected that Weightless networks will be deployed by network operators and a service provided to companies interested in machine applications. This will be explained in more detail in the coming chapters.

1.3 Possible applications

One of the key design aims of Weightless is to be application-agnostic – that is to provide a platform on which as many machine applications as possible can be based. One shared platform across multiple applications is clearly much more economic than separate networks for each major application. Hence, Weightless has not been designed with any one specific application in mind. It seems likely that once the networks are deployed many hundreds or thousands of applications will emerge – just as 'apps' are developed for Apps Stores.

However, it is helpful to have possible applications in mind so that the requirements for these can be understood and the technology designed to meet as many of these requirements as possible. At the time of writing, key applications include:

- *Smart grid*. The ability to interact remotely with electricity, gas and water meters. Initially this might be to regularly collect meter readings but it could expand to remote demand management and monitoring the quality of the supply. For gas and water meters a key requirement is to be able to operate for a decade or more from a battery. For all meters excellent coverage deep inside buildings is required.

- *Automotive*. There are myriad roles for machine communications in cars including engine management data and upgrades, safety-related information and emergency calling.
- *Transport*. Outside of cars, machine communications in public transport can help with time-of-arrival information, tracking of assets and more.
- *Healthcare*. There is an immediate need to be able to remotely monitor individuals with particular healthcare needs, such as some diabetics and this is an area where many more monitoring and location functions could be envisaged.
- *Asset tracking*. Monitoring of goods and parcels as they move around the world.
- *Financial*. Replacement of cellular transmission in credit card terminals and similar.
- *Smart cities*. A very wide range of applications sourcing data from sensors and processing this to provide valuable information. Sensors might include temperature, parking space availability, traffic level or even whether garbage bins need emptying. Information might include pre-planning routes for snow ploughs or salting vehicles, provision of guidance to the nearest parking space for drivers with connected sat-nav systems and so on.

Experience suggests that many more applications will rapidly emerge once networks are available. Some of these applications are considered in more detail in Chapter 10.

1.4 Key requirements

While each application has slightly different requirements, a network that aims to support all of them needs to have the following features or characteristics:

- *Support of a large number of terminals.* It seems quite plausible that there could be 10 connected devices per person. Some suggest that this may rise towards 100 devices per person. At 10 per person this implies approximately 10 times as many devices as mobile phones, which themselves are typically distributed across 2–4 national networks. A typical cell in a Weightless network might have between 100 000 and 1 million devices within it and a national network could easily cover 1 billion devices. Networks need to be scaled to enable this.
- *Long battery life.* A subset of applications is unpowered and there will often not be a user nearby to recharge batteries. Ten-year lifetimes from one battery are needed in many cases.
- *Mobility.* A subset of applications has moving terminals which need to be supported as they move, potentially across national borders.
- *Low-cost equipment.* For most terminals the value of the sensor will be very low – often in the region of $10. Hence, the wireless chipset to be integrated into the terminal needs to be much less than this. The lower the price the more applications that can be enabled, costs of $2 per chip or less would appear to be necessary.
- *Low cost service.* Equally, the owners of terminals will typically only be prepared to pay a few $ a year for a network subscription. Hence, the network costs must be low and the marginal cost of each terminal very low.
- *Global availability.* Some applications will require global roaming. Others, like automotive, will require that one solution can be fitted into all vehicles regardless of their country of destination.
- *Ubiquity.* Excellent coverage, including within buildings, is needed.
- *Guaranteed delivery.* Some applications require certainty that messages have been delivered. This may also require strong authentication and encryption.
- *Broadcast messages.* This may be sending the same data to multiple terminals – e.g. a software update. Or it may be sending a common message to terminals – e.g. to reduce energy requirements temporarily.

The technology will need to work with the following characteristics of terminals and applications:

- *Small bursts of data.* Most machines send data packets of the order of 50 bytes. Networks must be able to transmit these with minimal overheads if network traffic levels are not to become excessive.
- *Sub-optimal terminals.* In many cases terminals will be small and low cost and will have poor-quality antenna and limited power supplies. Antenna alignment/polarisation cannot be assumed.
- *Stimulated transmission.* Some events, e.g. a power outage, might cause a large number of terminals to simultaneously send alert messages. The network needs to be able to accommodate and control the resultant peak in loading.

However, there are some requirements which are needed for personal communications but less so for machine communications. These include:

- *High data rates.* Machines rarely need data rates in excess of a few kbits/s.
- *Low delay.* Most applications can tolerate a few seconds of delay on message transmission.
- *Seamless handover.* Most applications do not need seamless handover and can tolerate a few seconds break in communication while contact is re-established with the new cell.

The impact that all of these has on system design will be explained in subsequent chapters. Weightless is designed to meet all of these criteria and to benefit from those areas where requirements can be relaxed. Note that Weightless is not intended as a viable alternative for personal communications – it is assumed that cellular systems address this market adequately.

1.5 Market size

The likely size of the machine market is important information for those planning on investing in delivering services. However, at this embryonic stage, any predictions are more guesswork than science. History has shown that we often over-predict the short-term impact and under-predict the longer term. This was certainly the case for cellular communications.

Our approach has been to estimate the likely number of devices per person, counting those people that already have a mobile phone of which there are around 5 billion worldwide. It might be envisaged that the average person could have:

- Around three smart meters. Each building will have electricity, gas and water metering. There are multiple inhabitants per building but equally business premises with no inhabitants.
- Around one car.
- Around 0.5 healthcare related sensors – some people will have none, others multiple.
- Around one smart city sensor per person assuming a range of sensors for temperature, parking spaces, etc.
- Around two asset tracking terminals per person.
- Many others in the home, on the person and in the wider area.

This suggests that a value of 10 terminals per person is perfectly plausible and that if there are applications not yet foreseen this could easily rise much higher. For the sake of being conservative, the number 10 is taken for the remainder of this analysis. These terminals could be rolled out in a few years, but some industries such as healthcare and automotive are historically slow adopters of new technologies so a 10 year deployment horizon may be appropriate.

For terminals this suggests an annual market of the order 5 billion chips. With target prices of around $2 per chip this is a $10 billion market. Many new terminals may be developed and deployed as well – we make no attempt to determine the size of such a downstream market.

In terms of networks, a global deployment of at least one network per country would be needed over the next five years or so. This might equate to around 200 networks each with an average of around 5000 base stations, leading to a total market for 1 million base stations, or 200 000 per year. If multiple competing networks were deployed this could be two to three times greater. If base stations cost $5000 each then the annual market would be around $1 billion.

Annual service revenue per terminal is very uncertain and likely to vary dramatically depending on the traffic volumes and quality-of-service requirements. If the average annual fee were $10 per device then by the time market maturity is achieved around a decade from now the annual service revenue would be of the order $500 billion. It may be that over time annual fees decrease but even if they fall to $1 per device per year this is still a $50 billion/year global market.

Many analyst reports are likely to be written that will forecast in a much more scientific and detailed manner the likely market size – this section is more for illustration. It clearly shows this has the potential to be an enormous market, perhaps one of the largest growth areas in the communications sector in the coming decade.

1.6 How machine communications could change our world

Not everything is about money and it is worth thinking about what a world with widespread machine communications would be like to live in. Cellular communications have dramatically changed the way we live our lives both at work and in leisure and it seems quite possible that a successful widespread deployment of machine communications could make an equally dramatic change. Broadly, it might be expected to 'make everything work better'.

Machine communications could be the key enabling factor behind smart grids. This might enable us to reduce power requirements, have fewer stand-by power stations and charge electric vehicles without causing overload problems to the supply network. It could be at the core of 'green' energy solutions that enable us to dramatically reduce our carbon footprint.

Smart cities could make the lives of commuters and residents much better. Transport systems could become more efficient, street lights could be repaired when needed, garbage bins emptied as required, congestion avoidance and routing to an empty car parking space be provided and so on. Costs could be reduced due to saving unnecessary activities or automating others and the impact of reduced congestion and greater efficiency would also extend to environmental benefits.

Healthcare monitoring could save lives and enable many to live at home rather than stay in hospital. As well as increasing quality of life this could also reduce the burden on healthcare systems.

These are just some examples – with many other applications envisaged much greater benefits could be expected.

A very important observation is that machine communications could make a difference to what are currently seen as the major challenges facing us today – an aging population and global warming. Indeed, without machine communications it is hard to see how these can be addressed effectively. It may not be an overstatement that Weightless provides the mechanisms to save the planet!

2 The need for a new standard

2.1 Machine communications does not yet have the necessary standard

The observation, noted in the last chapter, that there was massive potential in machine communications is not a new one. Over the last few decades many have noted that the installation of a wireless connection into myriad devices would bring a range of benefits. However, the market for machine communications to-date has been weak. There are some cars with embedded cellular modems and some relatively high-value items such as vending machines are equipped with cellular packet-data modems. But the market today is only a tiny fraction of the size it has long been predicted to grow to. This is predominantly due to the lack of a ubiquitous wireless standard that meets the needs of the vast majority of the machine market as set out in Chapter 1. There is no current wireless system that comes close to meeting all of these requirements.

Cellular technologies do provide sufficiently good coverage for some applications but the hardware costs can be $20 or more depending on the generation of cellular used and the subscription costs are often closer to $10 per month than $10 per year. Battery life cannot be extended much beyond a few months. Cellular networks are often ill-suited to the short message sizes in machine communications resulting in massive overheads associated with signalling in order to move terminals from passive to active states, report on status and more. So while cellular can capture a small percentage of the market which can tolerate the high costs and where devices have external power, it will not be able to meet the requirements of the 50 billion device market. Indeed, if it could, it would have done so already and there would be no further debate about the need for new standards.

There are many short-range technologies that come closer to the price points. These include Wi-Fi, Bluetooth, Zigbee and others. However, being short range these cannot provide the coverage needed for applications such as automotive, sensors, asset tracking, healthcare and many more. Instead, they are restricted to machines connected within the home or office environments. Even in these environments there may be good reasons why a wide-area solution is preferred. For example, an electricity supply company is unlikely to accept that their meter is only connected via e.g. Zigbee, into a home network, which in turn connects to the home broadband. Were the home owner to turn this network off, fail to renew their broadband subscription or even just change the password on their home router, then connectivity could be lost. Restoring it might require a visit from a technician with associated cost. Maintaining security across such a network might also be very difficult.

Finally, it is critical that the technology is an open global standard, rather than a proprietary technology. With a wide range of applications there will need to be a vibrant eco-system delivering chips, terminals, base stations, software and more. The manufacturer of a device such as a temperature sensor will need to be able to procure chips from multiple sources and to be sure that any of them will interoperate with any wireless network across the globe.

Without a wide-area machine communications network that meets all of the sector requirements it is unsurprising that forecasts for connected machines have consistently been optimistic.

2.2 Barriers to delivering a machine communications network

While the needs of the machine sector have long been understood, the key problem to date has been a lack of insight as to how they could be met. Ubiquitous coverage requires the deployment of a nationwide network, and the conventional wisdom has been that such networks are extremely expensive. For example, a UK-wide cellular network can readily cost $2 billion with costs of spectrum adding another $1–2 billion. With the machine market unproven, such investments were not justifiable and would result in an overall network cost that would not allow the sub $10/year subscription fees needed to meet requirements.

The key to unlocking this problem is free, plentiful, globally harmonised low-frequency spectrum. It needs to be free or at least very low cost, to keep the investment cost low. It needs to be plentiful to provide the capacity to service billions of devices. It needs to be globally harmonised in order to allow devices to roam across countries and to enable the economies of scale needed to deliver <$2 chipsets. It needs to be low frequency to enable good range from each base station and therefore a relatively small number of base stations to provide ubiquitous coverage. Unfortunately, low-frequency spectrum is in very high demand and so rarely becomes available in sizable quantity, is almost never globally harmonised and where even a few of these attributes hold true, is extremely expensive.

The lack of spectrum that meets all these requirements has meant that up until now the only option for wide-area machine communications has been to make use of existing networks, predominantly cellular.

2.3 White space as a key enabler

In the last year, a new option has emerged for spectrum access. This is the use of the 'white space' spectrum – the unused portions of the spectrum band in and around TV transmissions. We will explore white space in much more detail in the next chapter, for now it is sufficient to note that white space spectrum meets all of the requirements for machine communications. It is unlicensed and so access to it is free. It is plentiful with estimates of around 150 MHz of spectrum available in most locations – more than the entire 3G cellular frequency band. It has the potential to be globally harmonised since the same band is used for TV transmissions around the world. Finally, it is in the perfect low-frequency band which enables excellent propagation without needing inconveniently large antenna in the devices. This is a 'game changer'. Access to white space provides the key input needed to make the deployment of a wide-area machine network economically feasible.

However, white space is not without its issues. These are broadly regulation and interference.

Regulation for white space is still developing in many countries. However, it is clear that white space access will require devices that have the following characteristics:

- Relatively low output power. The FCC has specified 4W EIRP for base stations and 100 mW EIRP for terminals. These are an order of magnitude lower than cellular technologies.
- Stringent adjacent channel emissions. White space devices must not interfere with existing users of the spectrum, predominantly TVs. Hence, the energy that they transmit must remain almost entirely within the channels they are allowed to use. The FCC has specified that adjacent channel emission need to be 55 dB lower than in-band emission, a specification much tighter than most of today's wireless technologies.
- The need to frequently consult a database to gain channel allocation. Devices may need to rapidly vacate a channel if it is needed by a licensed user. They must consult a database to be informed as to the channels they can use and must quickly move off these channels as required.

Interference can be problematic in white space. Many channels have residual signals from TV transmissions. These can be in-band emissions from distant, powerful TV masts that are too weak for useful TV reception but still significantly above the noise floor. Alternatively, they can be adjacent channel emissions from nearby TV transmitters some of which are transmitting in excess of 100 kW. In addition, since the band is unlicensed, other users might deploy equipment and transmit on the same channels as the machine network, causing local interference problems.

These are not insurmountable issues. But no current technology has been designed to operate in such an environment and so would be suboptimal at best. For example, an optimised technology could access around 90 MHz of white space after all the interference issues are taken into account, whereas an existing technology such as Wi-Fi or WiMax could only access around 20 MHz.

So white space spectrum provides the key to unlock the machine network problem. But it comes at the cost of needing to design a new standard.

2.4 Design rules for a machine communication solution

While the use of white space provides the need for a new standard, there are many benefits from designing a standard specifically for machine communications. Machines are very different from people as was explained in the previous chapter. Taking advantages of these differences allows the design of a system that is much more efficient, providing greater capacity than would otherwise be the case and hence having low cost. The predictability of most communications allows a very high level of scheduled communications as opposed to unscheduled, or contended, communications. The difference is akin to pre-booking passengers on flights so that each flight is full, but not over-crowded, rather than just letting passengers turn up, as with most trains, and suffering the crowding problems that occur. By telling terminals when their next communication is scheduled, future frames of information can be packed very efficiently and terminals can be sent to sleep for long periods, extending battery life.

Scheduling brings many other advantages. The first is efficiency. Contended access schemes can only operate up to about 35% channel usage – above this level the probability of access messages clashing becomes so high that very little information gets through. By comparison, scheduled access can achieve close to 100% efficiency. Scheduling can be enhanced by complex algorithms in the network. For example, these might prevent terminals close together in neighbouring cells receiving simultaneously. Or they can ensure that terminals suffering local interference are scheduled on frequency transmissions where interference is minimised.

Another design rule for machines is that coverage is typically more important than data rate. For example, it is more critical that all smart meters can be read than what the data rate of transmission is – as long as it is sufficient to transfer data regularly. In fact, most machine communications can be measured in bits/s rather than kbits/s or Mbits/s. As an example, a smart meter will typically send around 20–40 bytes of information perhaps once every 30 minutes. This equates to an average of 240 bits per 30 minutes or 8 bits/minute. There are applications that will require higher data rates, but speed is rarely critical. Hence, a good machine communications system design will trade off data rate against range. This can be achieved by spreading the data to be transmitted. Spreading involves multiplying the data by a pre-defined codeword such that one bit of transmitted data becomes multiple bits of codeword. The receiver can then use correlation to recover the codeword at lower signal levels than would otherwise be possible. Codewords are selected to have particular correlation properties and typically have length 2^n (e.g. 16, 32, 64). So, for example, multiplying the transmitted data by a codeword of 64 results in an improvement in link budget of some 18 dB but reduces the data rate by a factor of 64. Most buildings have a penetration loss for signals entering them of around 15 dB so spreading by this factor would provide indoor coverage to machines where only outdoor coverage previously existed. Some wireless solutions have spreading factors extending as far as 8192, providing great range, but very low capacity. While they bring important benefits, large spreading factors add complexity to the system design since they extend the time duration of important system control messages that all devices must hear, which in turn requires long frame durations. These design decisions make machine communication networks radically different in many respects from cellular solutions.

Another design rule, at least at this embryonic stage of the market, is flexibility. It is far from clear what applications will emerge. Even the balance between uplink and downlink is unclear – for example smart meters will likely generate predominantly uplink traffic while software updates, perhaps for car engine management systems, will be large downlink messages. This suggests that systems should be time division duplex (TDD) in order that the balance between downlink and uplink can be changed dynamically.

The terminal should be made as simple as possible, keeping complexity within the network. This is contrary to the trend in cellular communications where handsets have been becoming ever more powerful and complex. There are two key reasons to keep terminals simple. The first is to keep the cost as low as possible – as mentioned earlier many applications require chips with costs of the order $1–$2. The second is to minimise power consumption for terminals that are expected to run off batteries for 10+ years. This means that, for example, complex multi-antenna solutions should be avoided and that terminals should not be expected to make complex calculations to decode their messages. Even an apparently simple decision, such as requiring a terminal to respond on the uplink of a frame where it receives a message on the downlink could require it to process the downlink message much more rapidly, needing a more powerful processor. Careful design throughout is needed to achieve minimal terminal complexity.

Finally, there is likely to be an imbalance within a machine network where the base station has much more power and processing at its disposal and so can have a greater range than the terminals. This is of no value since the terminals need to be able to signal back and so the link budget must be balanced. With base stations transmitting often around 4 W (36 dBm) but battery powered terminals restricted to 40 mW (16 dBm) there is a 20 dB difference. This can be balanced by a combination of using narrower bandwidths in the uplink and using greater spreading factors, although this does come at a cost of longer transmission times with resultant increased power consumption. In general, battery life is longer where short, high-peak loading is avoided so this approach of lower power over a longer time is preferred.

Designing optimal technologies and networks for machine-based solutions does not require any technological breakthrough. But it does require great care in understanding the implications of each decision and it needs a system design that is radically different from a cellular network, with design decisions often appearing contrary to the conventional wisdom of the day.

2.5 The Weightless Special Interest Group

No machine communication system can become successful without being an open global standard. Indeed, history has shown that the only wireless communication systems that become successful are those that are open and widely supported standards. For this reason the Weightless Special Interest Group (SIG) was formed as a vehicle to develop the Weightless standard. The Weightless SIG was modelled on the successful Bluetooth SIG. Instead of making use of an existing standards body such as ETSI or the IEEE a new standards-making body was formed. This was because existing bodies had tended to become slow and bureaucratic. The Weightless SIG was launched in late 2011 using the *standard* version 0.6 as supplied by Neul, the inventors of Weightless. The aim of the SIG was to complete the standard to a level that suppliers could build interoperable equipment by the end of 2012. At the time of writing the SIG is planning its initial events and forming its subgroups. The intention is to segment the work into areas such as the physical layer, medium access control layer, applications, type approval, regulation and security.

One of the key principles of the SIG was that if the target price points are to be achieved for terminal devices then there is little room for royalty payments for intellectual property. Hence, any standard must be royalty free (the so-called Fair, Reasonable And Non-Discriminatory licensing with Zero royalties or FRAND-Z) for terminals. Weightless is not necessarily royalty free for base stations, with the final position to be adopted here still to be determined at the time of writing.

In more detail, all those who sign up to be a member of the Weightless SIG, whatever membership grade they opt for, are required to sign an IPR agreement relating to the terminals. This agreement broadly says that they will agree not to assert any IPR they may hold, or may develop while a member of the SIG, against any other SIG member who uses that IPR in relation to terminal development. In return, they are able to develop terminals or related items (such as chipsets) without needing to pay royalties to any other SIG member. There remains a possibility of a non-SIG member asserting IPR that they hold against those developing Weightless terminals. This cannot be prevented and, should it arise, will need to be handled in an appropriate manner. For example, SIG members might decide that the SIG should negotiate with any IPR holder on behalf of the collective membership. Of course, the more members of the SIG that there are, the lower the possibility will be of a non-SIG member asserting IPR. Note that members are not required to 'give up' any IPR.

They continue to hold all their IPR and can assert it against any non-SIG member, or against any SIG member in areas other than those related to Weightless terminals.

Regarding base stations there is no zero royalty position. SIG members and others are able to assert any IPR they hold against anybody making base stations. SIG members agree that they will license any relevant IPR they hold on a FRAND basis but do not need to declare this IPR or agree the detailed terms at any particular point. Non-members of the SIG, of course, can assert their IPR in any manner that they wish. At the time of writing, the SIG is considering whether to host a 'patent pool'. This would be a voluntary arrangement. Anyone with IPR that they felt to be relevant to Weightless base stations could declare this to the patent pool. Independent adjudicators would then consider all patents declared and assign a percentage value to each (such that they added to 100%) based upon how important they were to Weightless base stations. The patent holders and base station manufacturers would then agree on a reasonable total royalty fee to pay on each base station. The patent pool would collect these fees and distribute to members accordingly. Those who choose not to be part of the patent pool would have to assert their IPR independently and against each base station manufacturer.

The network is outside of the scope of the Weightless standard. There may be IPR relevant to network operations but this is not covered by the SIG and it will be up to those who hold the IPR, and those who might infringe it, to reach agreement outside of the SIG. However, should there be any critical IPR issues that threaten the deployment of the standard it is possible that the SIG might take some interest.

Another important function of the SIG is certification. The SIG will orchestrate type-approval and compatibility testing where it will check that terminals and base stations meet the Weightless standard (e.g. in areas such as sensitivity) and inter-operate with other Weightless equipment produced by different vendors. If equipment passes, it will be issued with certification enabling it to use the Weightless logo. At the time of writing, the certification process is still in the early stages of design. It was not expected to be needed until the standard was sufficiently stable for tests to be devised and manufacturers to start their equipment production process.

More information about the SIG can be found at www.weightless.org.

2.6 Summary

The value in machines having wireless communications has long been understood and a large market predicted for many years. That this has not transpired has been because of the difficulty of meeting all the requirements within the constraints of the available radio spectrum. These constraints changed significantly with the advent of white space which provides near-perfect spectrum with free access. However, the combination of the unique and unusual nature of that access and the very different characteristics of machine traffic compared to human traffic means that using any existing standard is far from optimal. Hence, the need for a standard designed specifically for machine communications within white space.

3 Working in white space spectrum

3.1 Defining white space

Most of the spectrum below 10 GHz across most of the world is allocated for particular applications and assigned to certain users. These include broadcasters, mobile phone network operators, defence departments and many, many more. The net result is that there is little obviously spare spectrum, particularly in the preferred frequency bands between around 300 MHz and 3 GHz where propagation is favourable but antennas are conveniently small. However, measurements of the actual utilisation of this assigned spectrum suggest that it is typically only used in around 20% of the locations. Such measurements undoubtedly underestimate usage but nevertheless it is clear that there is some potential for more efficient use of the spectrum.

One way to visualise where there might be underused spectrum is to plot on a map the strength of signal from the licensed user of that band. If colours are used to represent signal strength then those parts with no coverage at a given frequency will be uncoloured and appear white on a black and white map. Hence the term 'white space' for areas where there is potential for others to use the spectrum.

The most extensive surveys and models have typically been in the UHF TV bands where around 30%–50% of the spectrum appears to be unused in most locations. This UHF band has been the focus of work for a number of reasons including:

- There appears to be a relatively large amount of white space.
- The low frequency of the UHF band results in favourable propagation.
- Licences are often provided on a single transmitter basis rather than across the entire country. Therefore, licensed users do not own the white space, unlike for example the cellular bands where operators

typically have nationwide licences. This means that no negotiation is required with the licensed user to secure access; instead the regulator can legislate to allow it.

• Licensed use is relatively static. TV transmitters are seldom moved. It was also originally thought that the transmissions might be easy to detect although subsequent work has cast doubt on this.

White space occurs in the TV bands because frequencies can only be reused some distance apart. So if a particular frequency, or TV channel, is used at one mast it may provide coverage extending 40 miles or more, but its signal will still be strong enough to cause interference up to twice that distance. Hence, a geographical 'guard band' of some 40 miles or so, depending on terrain and other factors, is needed between the edge of one coverage area and the edge of another. If a high-powered TV transmitter were deployed in this guard band then interference with the existing transmitters would occur. However, a low-power, low-height transmitter might only have a range of some 5–10 miles. This could be potentially be deployed in the guard band without causing interference. It is, effectively, the disparity between the high-tower, high-power TV transmitter and the low-tower, low-power 'white space transmitter' that allows its operation.

A plan of the UHF TV bands in the UK is provided in Figure 3.1. Most European countries follow a similar plan.

3.2 Determining where the white space is

The original proposals for white space access envisaged that the devices would autonomously determine the location of the available spectrum by scanning the spectrum. Where no transmissions were detected, the device might assume that the spectrum was unused.

However, there is a problem with this approach often referred to as the 'hidden node problem'. This is illustrated in Figure 3.2. Essentially, the problem is that the white space device may not be able to detect a signal because of local topography or a relatively poor antenna and hence transmit in error, causing interference.

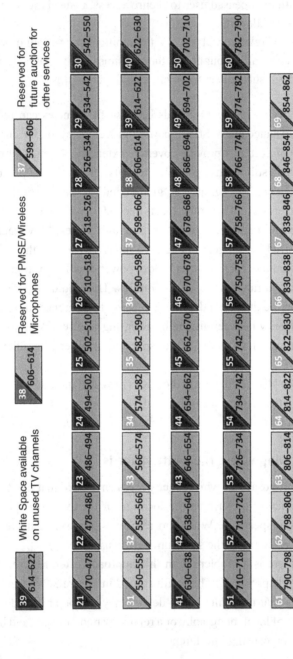

Figure 3.1 Plan of the 'white space' band in the UK [Source: Neul adapted from Ofcom publications]

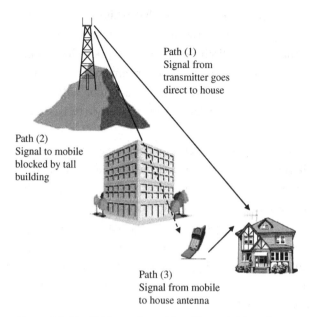

Path (1)
Signal from
transmitter goes
direct to house

Path (2)
Signal to mobile
blocked by tall
building

Path (3)
Signal from mobile
to house antenna

Figure 3.2 The hidden node problem [Source: Ofcom]

Resolving this problem either requires the primary user of the spectrum to accept occasional interference or for the white space device to be sufficiently sensitive that it can detect the signal even when the topography is problematic.

In general, primary users are not inclined to accept interference. For example, TV broadcasters argue that their viewers will not tolerate occasional loss of signal while radio microphone operators, the other users of the TV bands, point out the highly problematic consequences of a radio microphone suffering interference during a major stage show. There may be some users who are less concerned about interference but these have not emerged to date.

The alternative is to make the white space device sufficiently sensitive that the hidden terminal problem occurs so rarely as to be considered resolved. In the USA and UK substantial work has been undertaken to determine what sensitivity might be needed. Definitively determining sensitivity is not possible because it depends on real-world geometries, obstacles and topographies but substantial measurement and modelling have provided good guidance. The answer is that white space devices need to be extraordinarily sensitive, so much so that it is unclear whether such devices can actually be realised, let alone at a price that would appeal to consumers. It is possible that novel algorithms might be found in the future that can measure well below the noise floor but for the moment most accept that white space devices are unlikely to be based on detection.

If the white space device cannot determine accurately the spectrum available then an alternative is for a central infrastructure to make this determination and transmit the information to the device. In such an approach a white space device determines its location, perhaps using GPS, and then reports this, perhaps via a cellular data channel, to a central database. The database returns information on the frequencies available in the vicinity, possibly including additional information such as the transmit power that the device might employ or the likely duration that the channel will be available.

For such an approach to work the database must be supplied with detailed information about the licensed use including transmitter locations, transmitter power, antenna orientation and so on. It must also be aware of the operating parameters of the licensed systems such as the carrier-to-interference (C/I) level that the service needs to operate successfully. With this information it can derive a model of possible receiver locations and signal strengths and then, for every given point on a map it can determine what signal strength a white space device could transmit at without causing interference to the licensed service.

While this resolves the problems with detection, it requires considerable organisation and agreement. One or more databases need to be created and maintained and protocols designed for communicating with them. The white space devices now need to have a means of determining their location and also of communicating to the database without using white space (e.g. a wired link or a cellular link). In addition, licensed users need to update the database whenever their use changes, which can be frequently in the case of uses such as wireless microphones and cameras. This approach, known as 'geo-location', is now being adopted around the world.

3.3 The 'greyness' of white space

The previous discussion has been focused on how white space devices can operate without causing interference to licensed users of the spectrum. But it turns out that it is also important to consider the opposite case of licensed users causing interference to white space devices. In fact, one of the reasons that spectrum can appear unused is because the licensed users are leaving a guard band between two transmitters that would otherwise interfere with each other. Effectively, they have decided not to use the spectrum because of the interference that might occur. So a white space user can expect to frequently be attempting to use spectrum on which there are residual signals from licensed users. The extent of the problem will depend on the licensed use. In the case of white space within the TV bands, interference from distant but powerful TV transmitters can be very significant, often some 10–20 dB above the noise floor as shown in Figure 3.3 where the noise floor is represented by the horizontal line at −106 dBm. White space devices will need to work around this. This figure is based on measurements made near Cambridge, UK, across each of the TV channels. In each channel, numbered 21 through 69 we predicted the signal level using a standard propagation model (the solid bar) and then measured the actual received level both with an antenna aligned with the horizontal polarisation of the transmitter (the top line across the bar) and with deliberately misaligned vertical polarisation (the bottom line across the bar). Above each bar is information on the distance away of the relevant TV transmitter. The figure shows a number of factors. Firstly, reassuringly, predictions align well with measurements. Secondly, the interference on all channels available for white space is significant, especially if polarisation discrimination is not possible. Thirdly, signals from TV transmitters over 100 km away can still cause interference.

Interference might also occur from other white space users. To date, most thinking about white space is that it should be unlicensed – that is there is no restriction on the number of white space devices that can operate within the spectrum so long as they abide by rules designed to prevent interference to licensed users. This leads to uncertainty for white space

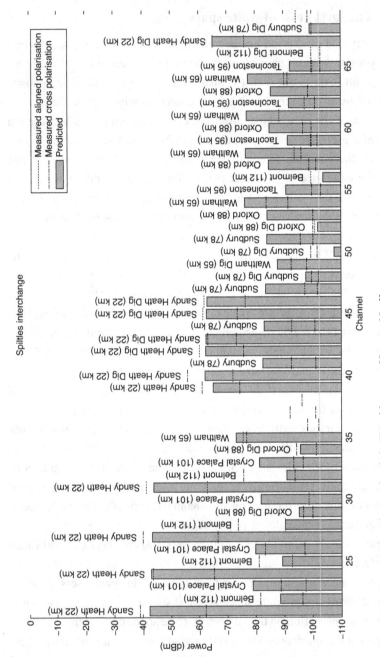

Figure 3.3 Measured interference levels in TV white space [Source: Neul]

devices as to how much spectrum they might be able to access both now and into the future. While bands such as 2.4 GHz have shown themselves able to accommodate very high levels of unlicensed and uncoordinated use, there are situations where Wi-Fi operation is significantly degraded or even impossible due to high levels of interference. Hybrid approaches are possible where spectrum is reserved as needed for white space, or the total number of users allowed at any one time restricted in some manner. These add complexity but could be subsequently introduced if problems with fully unlicensed operation emerge.

3.4 Design rules for white space

White space is unique spectrum. It is the first band where unlicensed users are allowed to mix with licensed users as long as they do not cause any interference to those users. This brings the benefit of free access to highly valuable spectrum but also a need to operate in an uncertain environment. Any system operating in white space should adhere to the following design rules.

Firstly it needs very low levels of out-of-band emissions. This minimises the interference caused to licensed users and so maximises spectrum availability. Achieving such low emission levels means that modulation schemes such as OFDM should be avoided as these tend to have relatively large adjacent channel emissions.

Next it needs to avoid interference caused by other unlicensed users which can be random and sporadic. Classic techniques for doing this include frequency hopping to rapidly move off compromised channels. However, hopping in a network requires central planning to avoid neighbouring cells using the same frequencies. Optimally planning the hopping sequences when different frequencies may be available in different cells and the sequence may need to dynamically adapt to interference is complex and requires new algorithms.

Where interference cannot be avoided the system needs to be able to continue to operate. Base stations can often experience significant interference from distant TV transmissions and require mechanisms such as interference cancellation to reduce its impact.

Finally, where there are few white space channels available, it can often be possible to increase availability by transmitting with lower power and hence causing less interference. Power control is therefore critical.

3.5 The US regulation

On 4 November 2008, the FCC published its rules on white space operation [1] in which it allowed unlicensed operation in the TV bands at locations where frequencies are not in use by licensed services. The FCC permitted both fixed and personal/portable unlicensed devices to operate in the TV bands. Fixed devices may operate at up to 4 watts EIRP (effective isotropic radiated power) on any channel between 2 and 51, except channels 3, 4 and 37, and subject to a number of other conditions such as a restriction against operation on the same channel (co-channel) as a TV station or on the first channel adjacent (adjacent channel) to such a station. Personal/portable devices may operate either as Mode I devices (operates only on channels identified by either a fixed or Mode II personal/portable device) or as Mode II devices (relies on geo-location and database access to determine available channels at its location). Personal portable devices may operate on any unoccupied channel between 21 and 51, except channel 37, and may use up to 100 milliwatts EIRP, except that operation on the first adjacent channels to TV stations are limited to 40 milliwatts EIRP. All devices (fixed and personal/portable) must include adaptive power control so that they use the minimum power necessary to accomplish communications.

One important detail in the specification is the requirement that power emissions are measured in a 100 kHz bandwidth rather than averaged across the entire 6 MHz TV band. This is intended to prevent multiple devices in the same area all using the full '6 MHz entitlement' of power but each in a narrow band within the same 6 MHz channel. If all these devices were close to a licensed receiver then it would be possible for the interference levels to be higher than allowed for in the modelling work. This measurement approach can be problematic for systems that use narrowband uplink channels in order to balance the link budget with relatively high-power base stations and low-power terminals – a preferred design option for Weightless as will be explained in subsequent chapters. The US restriction as currently set out would severely limit the transmit power of narrowband terminals using a narrowband uplink to around 2 dBm compared to preferred levels of around 15 dBm.

At the time of writing in 2011 many thought the US regulations inappropriate because:

- They can be circumvented by devices using CDMA where multiple devices all using the full 6 MHz entitlement could be located in the same area (and the resulting self-interference separated using the CDMA codes).
- In any interference situation, the interference is almost invariably dominated by the closest interferer as the interference received falls away very rapidly with distance. Further, in the case of TV reception, only interferers in the main beam of the receive antenna will have a material effect. The probability of multiple interferers being situated such that the combined interference is significantly more than a single interferer is very small indeed.

There are a myriad of further rules within the FCC specifications such as the requirement for devices to display the channel that they are operating on and any transmission of allowed channels from one device to another to be encrypted. These are important implementation details that we will consider in subsequent chapters. It seems possible that there may be changes to the US rules in due course to ease some of the restrictions that are most problematic to network operation in white space.

3.6 The UK regulation

The UK regulation as published by Ofcom [2] is based on detailed modelling of the potential interference caused. To understand the regulation it is worth understanding the interference issues in more detail.

Interference with licensed users can occur via two mechanisms:

1. Emissions from white space users that fall into the band used by licensed users. These are typically out-of-band emissions from white space devices that fall in-band for the licensed device.

2. Emissions from white space users that are outside of the band used by the licensed user, but which the licensed user's device is unable to filter adequately and hence results in interference. These are typically the in-band emissions from the white space device that fall in channels close to that used by the licensed device.

In practice, a combination of these two will occur. For example, a white space device operating two channels away from a TV receiver might cause interference as a result of its out-of-band emissions at $n + 2$ channels (where n is the channel the white space device is operating on). Simultaneously it might cause interference as a result of its in-band emissions being poorly filtered by a TV receiver with limited rejection at two-channel separation. Which of these is most significant will depend on the relative performance of the transmitter filter in the white space device and the receiver filter in the TV (or other licensed device). For a given device location, if a database knows (1) the possible location of licensed receivers (2) the frequencies they are using and their receive power levels at those frequencies (3) the performance of the licensed receivers and (4) the emission mask of the white space device transmitter, then the database can determine the maximum transmit power that the white space device can use before it causes interference.

Such an approach has been proposed by Ofcom in the UK. It has been approximated by the FCC in the US as described above where the emission masks of the white space devices have been fixed as has their maximum output powers. While simpler, the FCC approach lacks flexibility. For example, it does not allow for devices with poorer transmit masks which might be able to access less white space but be produced at lower cost. The implications of this are discussed further below.

The four pieces of knowledge that the database needs can be obtained as follows:

(1) The possible location of licensed receivers and receive power levels
These can be found through predicting the TV signal strength and establishing coverage contours showing the areas within which a TV

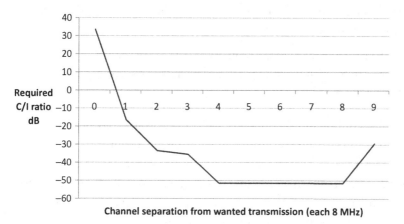

Figure 3.4 Protection ratios for TV [Source: Ofcom]

signal could be successfully received. They are generally calculated by broadcasters who have refined their models over many years of operation.

(2) The frequencies the licensed transmitters are using

These are equally well known and normally publicly available on broadcasters' websites or similar.

(3) The performance of the licensed receivers

This can be established either through using industry standards, such as the 'Blue Book' or alternatively, and more reliably, through measuring a range of available TV receivers and determining the levels of interference which cause visible signal degradation. Various margins can then be added to these, for example Ofcom suggest that any interference should be 10 dB below the level of degradation and that there should be a further 3 dB margin to allow for interference occurring both in-band and out-of-band. Taking these factors into account leads to the protection ratios shown in Figure 3.4.

This figure shows, for example, that, including the Ofcom 13 dB margin, 33 dB C/I is needed in band (i.e. the wanted TV signal must be at least 33 dB greater than the interference) but that at four channels separation the interferer can be up to 52 dB above the wanted TV signal. Note the

'$n + 9$' image frequency causes an increase in interference sensitivity nine channels away from the wanted TV signal.

(4) The emission mask of the white space device transmitter

This can be obtained from the device manufacturer, standards or measurements.

Once this information is complete, white space availability in a particular location can be found by using the following algorithm.

Draw a coverage contour around the white space device transmitter, operating at its specified transmit power level, out to the range where its transmitter power falls below the minimum TV signal level minus the specified in-band C/I protection margin (33 dB).

For every 'pixel' within this coverage area

{

    For every white space channel being tested for availability

    {

        For every white space channel from 9 below to 9 above the channel being tested for availability

        {

        Check whether the out-of-band emissions from the white space device transmitter will interfere with TV reception – if so remove the channel.

        Check whether the in-band emissions will not be sufficiently rejected by filtering in the TVs – if so remove the channel.

        }

    }

}

Any remaining white space channels can be used up to the specified white space device transmitter power level.

Clearly the number of channels returned can be improved through (i) reducing the white space device transmit power and so decreasing the range over which it can cause interference and (ii) reducing the

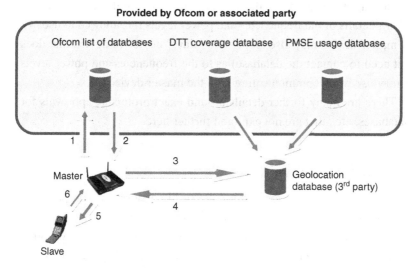

Figure 3.5 The proposed Ofcom process [Source: Ofcom]

white space device out-of-band emissions which reduces the signal levels falling in band to the TVs. Note that reduction in the out-of-band emission levels is only beneficial up to the point where the in-band emission from the white space device becomes the dominant interference case. So, for example, with the TV protection margin ratios assumed in Figure 3.4, once the white space device transmitter out-of-band emission levels in a channel one away from the carrier ('$n + 1$') fall below 50 dBc (33 minus -17) then interference will be dominated by the in-band signal from the white space interferer and further improvements in these out-of-band emissions will not deliver any further gains in white space availability. (The FCC rules are for adjacent channel emissions to be at least 55 dB down relative to the carrier – the analysis above suggests this is slightly over-strict and might be relaxed to 50 dB.)

Having understood how the database determines whether spectrum is available we now turn to look at the complete Ofcom process. This is shown in Figure 3.5.

Broadly, a 'master' white space device will first consult a list of databases provided on a website hosted by Ofcom (1 and 2). It will then select its preferred database from this list and send to it parameters describing its location and device attributes (3). The database will then return details of the frequencies and power levels it is allowed to use (4). A master device may also signal to a 'slave' device (a device that does not need to contact the database) as to the frequencies and power levels it may use when communicating with the master device (5).

There are many further details around exact protocols, approvals for databases, etc., that are not explored further here.

3.7 Other countries

At the time of writing, few other countries had significant proposals to use the white space. Where countries were starting to move forwards they were tending to adopt either the US or the UK approach. Adopting an existing approach is much preferred both to simplify the regulatory activities and to minimise the differences needed in equipment destined for multiple markets. This issue is explored further towards the end of this chapter.

3.8 Determining how much white space there is

The amount of white space can vary significantly from location to location. Firstly, the database will return differing amounts depending on the level of licensed use in the locality. Secondly, some of the returned channels may be unusable because of the level of interference from licensed users on the channel, or because of strong signals on adjacent channels.

Modelling all these different factors and determining white space availability is a complex process. A model starts with a map of the national TV coverage (here we use the UK as an example) based on the UK TV transmitter locations and uses a modified Hata model to predict the signal strength from each transmitter taking into account over-the-horizon propagation phenomena. It then models white space availability for a given white space device transmit power and white space device out-of-band emissions (OOB) characteristics. Finally, it rejects any channels where the technology will be unable to operate, either due to high co-channel interference from distant TV transmitters or strong signals on

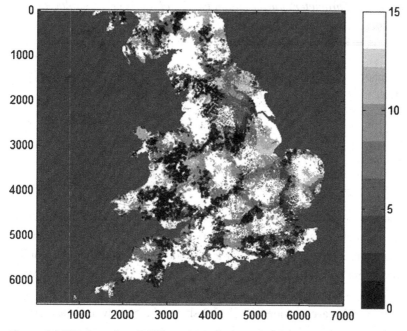

Figure 3.6 UK channel availability – Optimised technology [Source: Neul]

the neighbouring channel that the technology is unable to adequately reject.

A graphical representation is provided in Figure 3.6. Here the shading indicates the number of free channels, each one being 8 MHz wide.

The map illustrates the complexity of the situation. Broadly, more white space is available close to a TV transmitter than further away. This is because at the edge of coverage there are situations where viewers could select from two or three equidistant TV transmitters and hence transmissions from each must be protected. Also, in these regions, the received TV signal levels are low and hence the interference levels must be reduced to obtain the required C/I ratio.

While maps are visually striking, they do not allow for easy comparison of different scenarios – for this summary tables are more useful. In the tables below, results are provided for an optimised white space technology and a current technology (loosely based on WiMax and Wi-Fi), taking

Table 3.1 *Parameters for the 'optimised' and 'current' technologies [Source: Neul]*

Channel offset	Optimised: rejection (dB)	Optimised: OOB emissions (dBc)	Current: rejection (dB)	Current: OOB emissions (dBc)
0	18	N/A	0	N/A
1	45	55	20	28
2	65	65	40	50
3	80	80	60	70

into account all the factors discussed above. Tables 3.2 and 3.3 also detail the channel availability that would have been indicated by the white space geo-location database. The latter is as good as or better than the results for the individual technologies as it does not take blocking due to strong TV signals on neighbouring channels into account.

Two tables are provided. Table 3.2 shows the 50% white space availability (i.e. 50% of locations will have this amount or more white space available). Table 3.3 shows the percentage of points where no white space is available. The implications of this are different for peer-to-peer and networked technologies because a networked technology that has a lower transmit power level from the terminal device than from the base station may be able to site a base station just outside the area of non-availability and provide coverage into this area effectively at lower power levels. The assumed device parameters for the two technologies are in Table 3.1.

Beyond a three-channel offset, rejection is assumed to be perfect (i.e. any signal is completely filtered) and OOB emissions are assumed to be negligible. The optimised technology (based on Weightless) uses cross-polar operation and other techniques to deliver around 15–20 dB stronger rejection of TV signals than current technologies. It also has markedly improved transmitter filtering in order to achieve significant improvements in OOB emissions.

The tables provide results at four base station transmit (EIRP) power levels (terminal transmit power is assumed as 24 dB less than base station power taking relative power and antenna gains into account).

Table 3.2 *Comparison of 50% availability for different technologies [Source: Neul]*

50% TVWS availability level (MHz) at given EIRP	Optimised	Current	Database response to optimised
−5 dBW	120	24	184
0 dBW	112	16	152
5 dBW	96	16	136
10 dBW	88	8	120

Some of the key points to note regarding Table 3.2 (50% availability) are:

- As expected, lower transmit powers result in greater availability. Note this does not imply white space devices should be restricted to any particular power level, but that higher power operation may not be possible in some locations.
- The database returns many channels that the technologies are unable to use due to interference to the unlicensed device. For example, at 0 dBW the database indicates 152 MHz availability at 50% of locations, but the optimised technology can only access 112 MHz. This is due to blocking and interference from TV transmissions rendering many of the 'white' channels rather 'grey'.
- An optimised technology can access around four times the amount of spectrum of a non-optimised technology, suggesting that 're-banding' of existing technologies into white space would be sub-optimal.

Regarding Table 3.3 (no spectrum availability), points of note are:

- The optimised technology has spectrum availability close to that provided by a geo-location database that ignored the effect of licensed transmitter interference on white space devices. This is because non-availability is dominated by the need to protect weak TV signals. Hence, a technology with good OOB emissions characteristics can achieve availability close to that indicated by the database.

Table 3.3 *Comparison of locations with no availability for different technologies [Source: Neul]*

% locations with zero TVWS channels available	Optimised	Current	Database response to optimised
−5 dBW	1.1%	25.1%	1.0%
0 dBW	2.0%	29.2%	1.9%
5 dBW	3.2%	33.8%	3.1%
10 dBW	4.7%	43.0%	4.5%

Table 3.4 *Comparison of 50% availability for optimised technology [Source: Neul]*

50% TVWS availability level (MHz) at given EIRP	Optimised	Optimised, perfect rejection	Optimised, perfect rejection, no OOB emissions
−40 dBW	160	232	232
−30 dBW	152	200	200
−20 dBW	144	192	192
−10 dBW	128	192	192
0 dBW	112	152	152
10 dBW	88	120	120

- The current technology is much worse than the optimised technology with around 25%–40% of locations unavailable. This would be problematic for many of the applications considered for white space.

Further analysis is provided in Table 3.4 and Table 3.5 which focus on the optimised technology (Weightless).

These show how the white space availability changes with a wider range of transmit powers for the optimised technology, the optimised technology assuming perfect rejection, and the optimised technology with both perfect rejection and no OOB emissions (i.e. perfect OOB filtering). Note that the last column will be better than the database response provided in Table 3.2 and Table 3.3 as in those tables device

Table 3.5 *Comparison of locations with no availability for optimised technology [Source: Neul]*

% locations with zero TVWS channels available	Optimised	Optimised, perfect rejection	Optimised, perfect rejection, no OOB
−40 dBW	0.02%	0.00%	0.00%
−30 dBW	0.02%	0.00%	0.00%
−20 dBW	0.08%	0.06%	0.06%
−10 dBW	0.49%	0.45%	0.45%
0 dBW	2.00%	1.90%	1.90%
10 dBW	4.70%	4.50%	4.50%

OOB emissions are assumed whereas in the final columns of Table 3.4 and Table 3.5 no OOB emissions are assumed.

Table 3.4 and Table 3.5 show that:

- Perfect rejection makes a significant difference to the 50% availability, but improving the OOB emissions further makes no difference. This is because once the OOB ratios fall below the TV receiver's filter ratios then the in-band emissions from the white space device become dominant. Hence, any further improvement in white space devices should focus on rejection of interfering TV transmissions.
- Improving rejection or reducing OOB emissions makes little difference to the percentage of locations with no availability but reducing the transmit power levels has a major effect. Hence, if a white space device finds itself in an area of poor availability, its best option is to lower its transmit power. This further implies that regulation that allows variable transmit powers in white space will significantly enhance its usefulness.

The figures below help to illustrate why there are areas with no availability and how regulation might be changed to improve this. Figure 3.7 shows the predicted TV levels for the 32 white space channels in the UK (numbered sequentially rather than by actual channel number) at the four corners of a 1 km box drawn around a location with no availability:

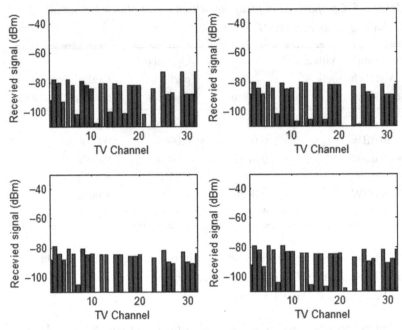

Figure 3.7 TV channel predictions around an area of no availability [Source: Neul]

It is immediately clear that this is an area of poor TV coverage. TV receivers need a minimum signal level of around −80 dBm and many of the signals here are only just above this level. There are no strong interferers. Hence, a lack of white space availability here is mostly to do with interference to TV signals rather than from TV transmissions interfering with white space devices. To investigate the problems further, it is helpful to isolate the channels that need protection. Clearly the only ones that need protection are the ones in this location that viewers are tuned to. When there is a dominant TV transmitter it is clear which these are. When there are multiple weak transmitters viewers could be tuned to any of the transmitters. This can often be seen in some locations where the rooftop TV Yagi antennas in a locality point in a range of directions. There is no 'correct' answer to which transmitters to protect; instead this is a regulatory decision balancing the protection to perhaps only one or two homes versus the loss of white space availability. A tractable way to solve this problem is to use a 'digital preferred service area'

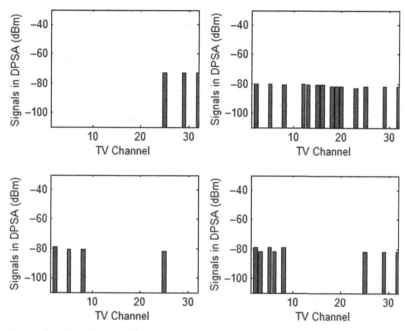

Figure 3.8 Viewable TV channel predictions around the same area as Figure 3.7 [Source: Neul]

(DPSA) margin. This allows for reception of channels up to a certain margin below the signal strength of the strongest predicted transmitter in the area. Channels falling below this margin are assumed not to need protection. Filtering Figure 3.7 with a DPSA margin of 3 dB gives the results shown in Figure 3.8.

The reason for non-availability becomes clear. The top-right location has a scattering of TV channels right across the white space band. Because these are at weak levels and TV receivers do not reject adjacent channels well then typically the channel and the $n \pm 1$ and $n \pm 2$ channels become unusable for white space operation. The spacing of these channels is such that this effectively prevents any white space operation.

If the channel usage was as in any of the other 'corners' then there would be substantial white space availability. A regulatory decision that only protected the transmitter providing these channels (each transmitter

Table 3.6 *Comparison of the 'Ofcom' and 'FCC' rules [Source: Neul]*

All at 5 dBW		Optimised	Current	Database (normal OOB, perfect rejection)
FCC rules	50% spectrum	80	16	96
	No spectrum	9.7%	35.2%	9.3%
Ofcom rules	50% spectrum	96	16	136
	No spectrum	3.2%	33.8%	3.1%

tends to transmit around 7 channels) would result in around 12 channels (96 MHz) of white space being available.

By comparison, in the US, operation is not allowed on a TV channel that is in use, and fixed devices are also not allowed to operate on the immediately adjacent channel. If those rules were applied to the environment modelled in this paper in the UK, the results would be as shown in Table 3.6.

The impact of the different rules is profound. Under the FCC rules the optimised technology has less spectrum available compared to the Ofcom rules and the number of locations with no availability triples. The current technology is less affected, predominantly because its performance is already poor. It is clearly advantageous, then, for regulators to use the more complex algorithm set out earlier, rather than the simpler approach of preventing use on adjacent channels for higher power operation.

3.9 Regulation in other countries

One of the key principles of Weightless is that it would be a global network, with service available across most countries and global economies of scale. For this to happen, white space needs to become widely available. Almost all countries use the UHF band for TV broadcasting in a similar manner (high-power transmitters, etc) so the white space exists everywhere. To become available for use, the regulator in each country needs to pass appropriate legislation, exempting white space devices from the need for a licence.

Each country is different and each regulator has their own preference between enabling innovative new services and risk-aversion. Historically, the UK and the US have often taken the lead in implementing new ideas and so it is no surprise that it is in these two countries that white space is becoming available first. Typically, once legislation is in place in a few countries others follow with similar regulation. Some, like Canada and Korea are often fast-followers, others may take some years. So it might be expected that white space would become available around the world over the coming years, with perhaps some additional countries having enabled it by the time this book goes to press and others taking as long as three years or so.

A relevant case study is that of ultra-wideband (UWB) which was also a technology seeking unlicensed access to licensed spectrum. The technology itself has not been a success but the regulatory process followed is instructive. UWB was initially proposed in the US and the FCC studied and consulted over a number of years. As the US process was drawing to a close, Ofcom also considered UWB and performed its own studies. Ofcom concluded that UWB would be better introduced at a European level and supported the European Commission in the development of pan-European regulation. This initially involved the EC requesting technical work from its associated technical specialist body, the Confederation of European Post and Telecommunications organisations (CEPT). The CEPT assessed the possibility for interference, paving the way for a proposal from the EC, very much in line with Ofcom's initial proposals. This was accepted by member states and became binding legislation, requiring all member states to implement rules allowing UWB within a six-month timescale. While this was on-going, a number of other countries such as Japan also conducted their own assessments and published appropriate regulation. The net effect of this was that UWB was enabled in the US initially, shortly followed by all of Europe and then within a year, much of the rest of the world. The regulations were not identical worldwide but there was much commonality.

White space access is following a very similar path. At the time of writing, there was much European activity. CEPT had been working on technical issues within a subgroup named SE43 for about two years and significant reports had been issued, broadly aligned with Ofcom's conclusions. At a political level the EC was frequently referring to the importance of white space to stimulate innovation, provide new services and assist in liberating additional spectrum. The spectrum policy group within the EC was giving consideration as to whether a pan-EC policy on white space should be developed but had not reached any conclusion. With clear European benefits from harmonisation of white space access including roaming and cross-border database alignment it seems likely that there will be EC activity to encourage or even mandate white space access in Europe, perhaps during 2012. There was also on-going activity in Canada, where a consultation had been issued, Japan, Korea, Singapore and others.

White space access is a relatively easy 'win' for a regulator. The regulator does not need to clear spectrum from an existing use or in any way compensate the incumbent. Access demonstrates the regulator's appetite to enable innovative services. If important white space services are deployed in other countries, such as Weightless networks that enable key smart-city societal objectives, then there will also be increasing pressure on a regulator to enable the same benefits in their country.

Hence, there is every reason to be optimistic that white space will be enabled around the globe in the next few years. The exact timing of how this will occur is difficult to predict, but likely there will be a few countries in 2012, with a rapid 'snowball effect' such that during 2013 and 2014 the rest of the world will follow suit.

3.10 Other spectrum

Weightless is not restricted to white space spectrum. Indeed, any wireless system can be moved to many different frequency bands, as has been shown with cellular systems. Operation outside of white space would generally be simpler as there would not be as many regulatory requirements such as database access and transmit power restrictions. Weightless could readily operate in such bands without any modifications apart from simple changes to the algorithms that determined base station frequency assignment. Weightless could operate in other unlicensed bands or in licensed spectrum.

At the time of writing there were some other possible bands that had been suggested but not investigated in detail including:

- white space access into VHF bands
- quasi-white space access into military bands
- use of licensed spectrum at 600 MHz
- shared use of emergency service bands around 380–450 MHz.

Some of the important factors when considering the use of different bands are:

- If the frequency is below around 400 MHz then terminal antennas become quite long which may not suit certain applications.
- If the frequency is above around 1 GHz then the signal will not propagate as far resulting in smaller cells or worse coverage.
- If the channel bandwidth is different from 6–8 MHz then significant equipment redesign may be needed.
- If the band is only in use in one country it may be difficult to get economies of scale on the manufacture of any bespoke equipment needed. Roaming to other countries may also be problematic.

One of the reasons to seek alternative bands may be concern over the interference that might result in white space. It is possible, although unlikely, that intensive use of white space by a wide range of applications might result in high levels of interference, reducing the capacity of a Weightless network, or in extreme making it impossible to operate. Experience of the 2.4 GHz 'Wi-Fi' unlicensed band which continues to work well despite very intensive use suggests that white space will not suffer from this issue. Further, the relative difficulty of using white space with the need for geo-location will tend to reduce the number of applications that seek to use it. Finally, the database access protocol can be modified to limit the usage of the band – this is a topic for further study with some regulators. However, for some applications such as smart grids, which might become part of the critical national infrastructure, greater certainty of access may be sought. This might be achieved either with exclusive spectrum or with prioritised shared access.

It is quite possible that there might be exclusive spectrum available within the bands covered by the core Weightless specification (470–790 MHz). For example, in the UK there are plans to auction the '600 MHz' band (TV channels 31–38) for exclusive licensed use. A Weightless network operator could purchase one or more channels of this spectrum to use in conjunction with other white space spectrum, or if they acquired four channels, to use as the only spectrum needed for the network. Whether this is likely to happen will depend on the success of the standard and the cost of the spectrum at auction. Other similar pieces of spectrum might become available in other countries. A key advantage of the 600 MHz band is that it is within the tuning range of standard Weightless equipment with 'standard' 8 MHz channelisation.

Another development might be white space access into other spectrum. For example, the military might deploy its own white space database to enable access to some of its spectrum and then offer this access for a fee to a limited number of applicants. A Weightless operator might reach agreement with the military to be an exclusive user of one of its bands. While access would not be certain because the military might reduce white space availability on certain occasions, historical data might suggest that these occasions are very limited.

To summarise, TV white space has many very desirable characteristics for Weightless and is the key target band. However, Weightless is not restricted to these bands, will work well in alternatives, and there may be operational reasons why a network operator is willing to pay for spectrum access in order to gain increased certainty of availability.

3.11 Conclusions

This chapter has introduced the concept of white space as 'gaps' in existing licensed spectrum usage. White space regulatory rules are still a work in progress, but to date regulators have opted for a geo-location approach where devices determine their location and request channel availability from a regulatory-approved database. The frequencies returned by this database depend on the rules adopted by the regulators and there are significant differences in the rules put forward in the UK and the US. White space devices may not be able to work on all channels returned due to interference from licensed use and modelling the resulting availability is complex but shows there is adequate white space spectrum in almost all locations. The modelling reveals the optimal parameters to adopt within a white space radio in order to maximise white space availability. These parameters have been used in designing Weightless as will be discussed in more detail in subsequent chapters.

References

[1] FCC 10–174, 23 September 2010.
[2] http://stakeholders.ofcom.org.uk/consultations/geolocation/

4 Weightless in overview

4.1 The key requirements

Weightless is designed to enable machine communications in white space spectrum. This leads to two sets of requirements – those related to machine applications and those related to operating in white space. The machine requirements were introduced in Chapter 1 and are summarised below:

- *Support of a large number of terminals*. A typical cell might have between 100 000 and 1 million devices within it and a national network could easily contain 1 billion devices.
- *Long battery life*. Ten year lifetimes from one battery are needed in many cases.
- *Mobility*. A subset of applications has moving terminals which need to be supported as they move, potentially across national borders.
- *Low-cost equipment*. Costs of $2 per chip or less would appear to be necessary.
- *Low-cost service*. Network costs must be low and the marginal cost of each terminal very low.
- *Global availability*. Some applications will require global roaming. Others, like automotive, will require that one solution can be fitted into all vehicles regardless of their country of destination.
- *Ubiquity*. Excellent coverage, including within buildings, is needed.
- *Guaranteed delivery*. Some applications require certainty that messages have been delivered. This may also require strong authentication and encryption.
- *Broadcast messages*.
- *Efficient transmission of small bursts of data*. Most machines send data packets of the order of 50 bytes.

- *Accommodating sub-optimal terminals.* In many cases terminals will be small and low cost and will have a poor-quality antenna and limited power supplies.
- *Event-stimulated loading peaks.* The network needs to be able to accommodate and control the resultant peak in loading.

As discussed in the previous chapter, white space operation leads to the following requirements:

- *Very low levels of out-of-band emissions.* This minimises interference caused to licensed users and so maximises spectrum availability.
- *Avoid interference caused by other unlicensed users.* This can be random and sporadic.
- *Reduce the impact of interference where it cannot be avoided.* Where interference cannot be avoided the system needs to be able to continue to operate.
- *Reduce power where there are few white space channels available.* It is often possible to increase availability by transmitting with lower power and hence causing less interference.

Some of these requirements have immediate design implications as discussed in the section below.

4.2 Immediate design implications

Each of the requirements has different implications and there is no obvious order to tackle them in. Perhaps the most important one is the need for ubiquitous coverage. This implies a solution with a cellular architecture. Along with this comes the need for a network, roaming, authentication, billing and many other aspects of cellular technology. It implies that at a high level the system architecture will look very similar to that of conventional cellular systems. However, as will be discussed, the scale of the various network components can be much reduced compared to cellular.

Achieving coverage even deep indoors has a further implication. Current cellular systems have relatively poor indoor coverage and white space transmitters will typically be restricted to lower power levels than cellular base stations. One solution would be smaller cells but the result of this would be a very costly network deployment. Instead, a way needs to be found to achieve better coverage than cellular with fewer base stations and less transmit power. The only way that this can be achieved is to use spreading. Direct sequence spread spectrum (DSSS) multiplies each transmitted symbol by a codeword resulting in either a high transmitted data rate or longer effective bit duration. This enables range to be extended at the cost of data rate. It is a technique employed in GPS transmissions to allow the weak satellite signal to be received with a handheld device at ground level. Spreading can achieve a 30 dB gain in link budget or more – sufficient to achieve the objectives set out above. However, it has other design ramifications, discussed in the next section.

The need for devices to work for years from batteries and the regulatory restrictions that result in lower power for the portable devices causes further problems. With more powerful base stations than terminals there is a risk of an unbalanced link budget where the terminals can hear the base station but not vice versa. In Weightless it is quite normal for the base station to be transmitting at 4 W EIRP (36 dBm) but the terminal to only transmit at 40 mW EIRP (16 dBm) resulting in a 20 dB difference in the link budget. This can be accommodated by using narrower bandwidth channels on the uplink resulting in a lower noise floor at the base station receiver and enabling the SNR targets to be achieved. Using uplink channels of 1/64th of the bandwidth of the downlink provides a noise floor 18 dB lower which approximately balances the budget. Such an approach works well with the UK regulations, but as mentioned previously, the US regulations measure transmit power in narrower bandwidth which mitigates against using narrow channels. The implications of this are described in more detail in Chapter 6.

Another implication of the use of white space is to adopt time division duplex (TDD). This is because the availability of two appropriately spaced white space channels as needed for FDD cannot be guaranteed. TDD also provides flexibility in that at the time of design it was far from clear what the balance of downlink versus uplink traffic would be on a machine network.

Another implication of white space for the initial design was a need to be able to avoid random interference from other unlicensed white space users. The classic approach to this, used by systems such as Bluetooth, is frequency hopping. Hopping also brings many other benefits such as averaging of self-interference, good neighbourly behaviour to other white space users and mitigation against being stuck in a fade.

White space operation also strongly biases designs towards structured synchronous solutions where there are frames, frame headers and devices are provided with allocations rather than transmitting randomly. This is because base stations must communicate information to terminals such as the frequency hopping pattern that is in use and in some cases restrictions on transmit power. With such a structure in place it then makes sense to schedule traffic rather than allow devices to transmit whenever they wish since scheduling gives much higher efficiency of loading by avoiding random transmissions colliding. This does require a somewhat more complex system and terminal design but still one significantly simpler than even 2G cellular systems.

4.3 Subsequent design thinking

These design decisions have subsequent impacts. One of the most far-reaching is the use of spreading. Spreading extends the duration of messages. Any frame header information transmitted in a cell must be at the highest spreading factor supported in the network to ensure that all terminals are able to receive it. Although every attempt has been made to minimise header information, it cannot be removed completely. The minimum size of the header information times the symbol rate times the maximum spreading factor dictates the time spent at the start of each frame transmitting header information. This works out at around 100 ms as will be discussed in more detail in subsequent chapters. In order to keep the overhead of the header information to below 10% this implies that the frame duration should be of the order 1 s or more. In Weightless the duration can be set as a variable within the network but a length of 1–2 s is recommended. This is much longer than the frame duration in most wireless systems, hence Weightless can be considered to have a long frame duration.

The long frame duration has implications. One is that the minimum round-trip delay is about the frame length – of the order 2 s – at best case and twice this at worst case. This would be disastrous for voice calls or even for Internet browsing but is typically not a problem for machines. The second is that it allows a different base station implementation where most of the processing is removed to the core network. There is ample time for the core to prepare a complete frame and send it to the base station for conversion to RF and transmission. This enables low-cost base stations, a simple upgrade path and more intelligent scheduling decisions across the network.

Another set of implications flow from the requirements for a long battery life. This implies terminals that want to save energy are able to enter into a sleep mode. However, too long a sleep mode would compromise the ability to contact them unexpectedly (e.g. with an alert message) or increase the probability that network changes such as updated frequency assignments would take place while asleep. For Weightless, calculations suggest that a sleep time of around 15 minutes would result in a battery drain sufficiently small that battery life is constrained more by the shelf life of the battery than the current consumption. Weightless has therefore been designed with the idea of a super-frame that repeats at around 15-minute intervals. The start of a super-frame is a point where all terminals are expected to wake up and listen and hence it can be used to alert them to network changes and send other relevant control information.

Of course, low battery drain is only achieved if terminals listen for the minimum amount of time then revert to sleep mode. To achieve this all the information needed by a terminal is contained within the header of each frame. Hence, any terminal need only listen for about 100 ms and if there is no information destined for it, can then return to sleep. This requires careful header design to avoid the terminal having to listen to subsequent frames to obtain a complete set of information. For example, it implies that the hopping sequence cannot be communicated by listing in the header the next frequency to be used and requiring the terminal to listen to sequential frames until the pattern repeats. Instead, the entire pattern, albeit efficiently encoded, must be transmitted in each frame.

The need to be able to handle sudden peaks in traffic due to some event such as a power failure stimulating multiple devices requires careful control of the uplink resource. Mechanisms to forestall devices sending alerts once the error condition has been noted by the network are also needed.

Mobility support requires terminals to be able to move from cell to cell. In cellular systems the network controls handover based on measurement reports provided by terminals. However, this generates substantial network traffic in terms of measurements and imposes a heavy battery load on the terminals when monitoring adjacent cells. Because machines do not need seamless handover a much simpler approach is adopted in Weightless. Handover is almost entirely driven by terminals (the exceptions to this will be discussed in subsequent chapters). Once a terminal detects it has moved out of coverage of a cell it re-starts its acquisition process and attaches to a new cell providing coverage. This means there is little need for any signalling traffic either from the terminal or the network which dramatically improves network efficiency. Hence, handover is terminal-driven.

The need to achieve stringent adjacent channel emissions has an impact on the modulation approach used. Tightly filtering OFDM transmissions tends to distort the waveform more than the same degree of filtering on single-carrier modulation due to the higher peak-to-average power ratio requirements of OFDM. Hence single-carrier modulation is preferred for white space operation. Weightless uses single carrier modulation but benefits from the frequency domain equalisation possible in OFDM by using single carrier frequency domain equalisation (FDE) where a cyclic prefix is inserted as in OFDM and then used to determine the channel frequency response.

Finally, the need to handle a very large number of devices requires considerable intelligence in the network to schedule communications and adapt network parameters according to load. The loading problem is exacerbated by the varying nature of the frequency resource available

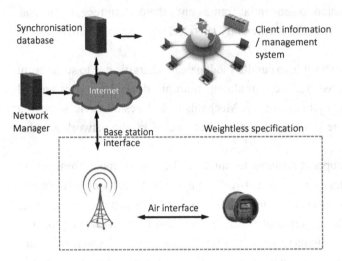

Figure 4.1 Overview of Weightless

with white space channel availability changing and interference potentially occurring randomly from other white space users.

This, then, sets the key parameters of Weightless as a TDD system with single-carrier modulation, direct sequence spreading, broadband downlink and narrowband uplink, long frame duration, frequency hopping at the frame rate and 15-minute sleep cycle capability.

4.4 System overview

A high level overview of Weightless is shown in Figure 4.1.

Terminals signal to base stations over the air interface. Base stations send and receive frames of information into the core network via a backhaul connection which might be routed through the Internet or through private connections. The Weightless specification covers the air interface and the base station interface allowing multiple companies to develop terminals and base stations. The core network functionality resides within a network manager, which may itself be a virtual entity within the cloud. This delivers frames of information to the base stations as well as coordinating frequency hopping assignments, managing location records and more. Information sent by the terminals is then routed to a

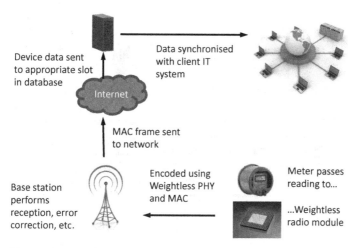

Figure 4.2 Overview of information flow within a Weightless network

'synchronisation database' which acts as an interface between the Weightless network and any software system the client might be using such as SAP or Oracle.

Another way to view this is to look at the information flow as shown in Figure 4.2.

This shows a terminal such as a smart meter passing a reading to an inbuilt Weightless module. This encodes and transmits it over the air interface to the base station which performs functions such as error correction before forwarding the frame to the core network. This routes the information within a frame to the appropriate client interface function.

A third way to look at the network is in a functional layered diagram as shown in Figure 4.3.

At the highest level the database communicates to the client IT system using a layer that is likely specific to that client based on their preferred IT solution. Below that the application in the terminal communicates with an application layer within the database. This allows application specific coding to be implemented. The terminal communicates with a Weightless radio through an interface to the radio unit and the radio then uses MAC and PHY layers to communicate to the base station. The base station sends frame-level information onto the database. Note

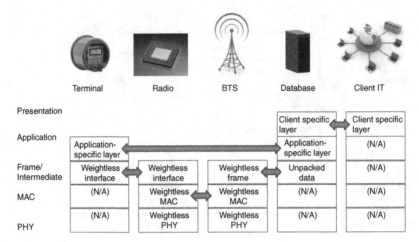

Figure 4.3 Layered diagram of a Weightless network

that the only interfaces defined within the Weightless specification are the MAC/PHY level air interface and the frame-level base station to database interface. Application-level specifications might be provided but are likely to be developed by application groups outside of Weightless and the client-specific layer interface does not need standardisation as it is likely to be somewhat bespoke.

Subsequent chapters will consider each of these different layers and interfaces in more detail with applications being considered in Chapter 10.

5   The network

5.1 Introduction

It would be possible to write a book about Weightless without much discussion about the network. This is partly because the network is not included within the specification and partly because the network is relatively simple.

The Weightless specification covers the interface from the base station into the network. However, unlike cellular networks there are no defined interfaces within the network. This is because it is envisaged that all the network functions can be run on a virtual machine within the cloud. The interface from this machine to the base stations is well defined. The interface from this machine to the client networks will likely be bespoke for each client network and therefore not appropriate for specification. Hence, there are no further interfaces to standardise.

It might seem rather incredible that the core network of a cellular system consisting of many large and expensive computing platforms can be collapsed into a single virtual machine within Weightless when a machine network is likely to have many more subscribers than a cellular network and further much of the intelligence typically found in a base station in a cellular network has been transferred to the core. But a machine network has a number of simplifications over a cellular network including:

- All traffic is genuinely packet based. Hence a fully IP-based switching solution can be used without any need to support time-critical applications such as voice over IP.
- A Weightless network will typically only have around 1/4 of the number of base stations of a cellular network.
- Seamless handover does not need to be supported.

- With longer latency, there is much more time to process data allowing for lower-performance platforms.
- Cellular systems support a wide range of services such as voicemail, short code dialling and much more that is not required for machines.
- Overall traffic volumes are much smaller. While a 4G base station might generate 100 Mbits/s of traffic or more, a Weightless base station will typically generate 100 kbits/s on average – some three orders of magnitude less.
- Cloud-based computing platforms are becoming rapidly more powerful. It might be that a re-design of a cellular network would result in some of the core functions being moved to the cloud.

Cloud-based hosting of the core network is extremely advantageous. It means there is no initial outlay for the core network, the processing power can grow as needed and backup functionality/quality of service can be provided by the cloud supplier rather than through deploying redundant solutions. Upgrading to new functionality is simple.

The core network does contain both classical functional elements such as location registers, and also algorithmic functions such as the frequency hopping assignment. In this chapter we take a brief look at these elements as this helps understand how the PHY and MAC work and why, for example, the frequency hopping list that is presented to terminals is structured in a particular manner.

5.2 Functional elements

The core network needs to perform the functions listed below. In a classic cellular network each of these functions might be a separate machine or database with a defined interface between them. In Weightless they will be part of a software suite running on a virtual machine.

- *Base station controller*. This provides the information to the base stations for transmissions in pre-formatted frames and provides a point of connection into the network for multiple base stations.
- *Billing system*. This records utilisation levels and generates appropriate billing data.

- *Authentication system.* This holds terminal and base station authentication information.
- *Location register.* This retains the last known location of the terminals.
- *Broadcast register.* This retains information on group membership and can be used to store and process acknowledgements to broadcast messages.
- *Operations and maintenance centre (OMC).* This monitors the function of the network and raises alarms when errors are detected.

In addition, there is a need to access a white space database providing information on the available white space spectrum in the area. At present, it seems likely that this will be a service provided by third parties to users of white space spectrum and not part of the core Weightless network. However, if needed it could be built into the core network.

Exactly what each of these functions do is very much up to the network operator. Some general guidance is provided below.

5.2.1 Billing system

Weightless networks may offer a very wide range of billing arrangements from a single cost to clients, through aggregate traffic costs to detailed billing on a per message basis (although detailed billing is thought unlikely to be widely used given the low value of most machine messages). The level of data gathered is up to the operator. However, it may be that certain roaming partners require more detailed usage information so if this is not gathered there may be some restrictions to roaming.

5.2.2 Authentication and encryption

In outline, the role of the authentication and encryption system is to encrypt the data received from the clients and authenticate and decrypt the data received from the terminals. The encryption system holds the keys to enable this.

Keys are entered into terminals at manufacturing time or deployment using a pre-agreed database of keys and terminal identities provided by the Weightless SIG. Sets of identities and keys might be sent periodically to each manufacturer as they exhaust the previous list and are entered into the master system database to be distributed to local networks as needed.

More details of the way that authentication and encryption work within the Weightless system are provided in Section 8.1.

5.2.3 Location register

In a cellular network the location register records the cell, or paging area within which a mobile is located such that it can be paged when it has an incoming message. This function is also required in Weightless. The location register can go a step further and store a more precise location. With many terminals being static, such as smart meters, there can be numerous advantages to knowing where these are. These advantages include being able to schedule communications on frequencies free of interference in the vicinity and ensuring multiple closely located terminals do not transmit simultaneously. The precise location could come from a location report provided by the terminal itself, based for example on GPS location. Alternatively, Weightless offers the ability to establish terminal location using one of two different forms of triangulation as discussed in Section 8.3. When this function is invoked the location register can collect timing reports from base stations and determine by trigonometry the location of the terminal.

5.2.4 Broadcast register

This retains information on group membership and can be used to store and process acknowledgements to broadcast messages. Clients can send messages to it to update group membership and terminals can request current group membership status from the register.

When a broadcast message is initiated the broadcast register provides the list of terminals to be included. This is sent to the location register to determine the list of affected base stations before the message is passed to the appropriate base stations.

5.2.5 Base station controller

The base station controller provides a single point of communications to the base stations and then distributes the information received to other network elements as needed.

A core function is the assembly of frames of information that the base stations will transmit. To do this it needs to make scheduling decisions.

The base station controller is also responsible for planning radio-related resources. These include frequency planning, code assignment, synchronisation word planning and load balancing as discussed below.

5.3 Frequency assignment

The frequency planning problem is relatively complex in a Weightless network. Ideally neighbouring cells should operate on different frequencies. In the downlink it is possible that terminals on the edge of one cell could receive interference from transmissions in the neighbouring cells. In the uplink there is a serious risk of base station to base station interference during misaligned TDD frames. This is because different base stations may choose different TDD splits according to the traffic mix in their cells. As a result one base station might be transmitting while a neighbour is receiving. Propagation from one base station to another is often relatively good as both can be in elevated positions and there may even be line-of-sight between them. A base station attempting to receive while a neighbour transmitted could easily be deafened by the neighbour's transmission. Because there is typically alignment of the start of each frame between base stations, this interference will only take place around the middle of the frame, but nevertheless it could have a marked effect on performance. As a result, while not essential, the use of different frequencies in neighbouring cells is strongly preferred.

Each cell will have a set of frequencies available returned by the white space database to the frequency planning function in the core network. In neighbouring cells it is likely that the returned set will be similar, but there may be some differences. This means that neighbouring cells may not have the same number of frequencies available to them and that the lists of frequencies may have significant overlap but not be identical. These lists may change hourly as the licensed use, particularly from wireless microphones, changes. Further, some of these frequencies may be compromised by interference from licensed use or from unlicensed users in the vicinity. As a result of other unlicensed use it may be necessary to remove a particular frequency from the available list in a particular cell for a limited period. Hence, the frequencies available form a complex and shifting set from which an optimal assignment needs to be found. Quite how this is done is up to different providers of core network solutions. However, in designing the Weightless MAC we had to understand this problem better so that the list of hopping frequencies in a cell could be optimally encoded.

In principle, given a set of frequencies in each cell and understanding which cells are adjacent, an exhaustive search could be conducted for the assignment that minimised the number of hops where neighbouring cells used the same frequencies. However, this rapidly becomes intractable. For example, with 8 cells each with 8 frequencies available to them there are $8! = 40\,320$ ways of arranging the frequencies in each cell and $40\,320^8$ total combinations – around 7×10^{36} combinations! This is well beyond the computing power of all systems. With a complete Weightless network having more than 5000 cells clearly a different solution is needed!

Some detailed modelling showed that if the list of hopping sequences allowed in a cell was restricted to incrementally increasing (so 1,2,3,4 is allowed as is 2,3,4,1 but not 2,1,4,3) then the simplified solution is very close in performance to the optimal solution. Indeed, in many cases, it is just as good. But the number of options available in a cell falls from $n!$, where n is the number of frequencies, to n. So the number of combinations for 8 cells with 8 frequencies now becomes 8^8 – around 16 million. This is perfectly tractable but again a problem occurs when extended to 5000 cells. Suffice to say here that there are solutions to this problem involving planning discrete overlapping areas.

The main outcome for the Weightless standard is that the hopping list can be assumed to be sequential which assists in encoding it within the

MAC layer and might be exploited by terminals when seeking to acquire the network.

One complexity in frequency planning is the manner in which the system reacts to random interference, typically coming from other unlicensed users of the spectrum. Take the case of an interferer within a part of one cell. The base station will notice that on a particular frequency, f, on its hopping sequence the error rate associated with its transmissions will suddenly increase for some of its terminals and it can deduce this is due to interference. To some extent it may be able to schedule communications to terminals in the vicinity on different frequencies and so minimise the impact if it has sufficient information about terminal location and they are mostly static. In this case, it might decide to leave the hopping sequence unchanged and use the scheduling algorithm (described below) to work around the problem. Alternatively it might decide that the interference is too widespread or it has insufficient information to work around the interference and that it wishes to remove the frequency f from its hopping sequence. This requires a complete frequency re-plan, in principle across the entire network. To understand this, imagine the cell in question, cell 1, was using frequencies 1,2,3,4 and a neighbour was assigned 2,3,4,1. There would be no interference between them. Now if cell 1 removes frequency 4 its hopping pattern would become 1,2,3,1,2,3,1 . . . This would cause interference with cell 2 using 2,3,4,**1,2,3**,4,1 on the frequencies highlighted in bold. But if cell 2's frequencies are changed this will cause interference with its neighbours and so on.

In practice, it is not recommended to change the hopping assignment more often than the super-frame duration. This is because the network uses the broadcast frame at the start of the super-frame to inform terminals of a change in the hopping pattern. If a change is made more frequently, terminals that only listen to the broadcast channel (e.g. because they are battery powered) will not be alerted and will fail to find the network when they awake. This will cause them to reacquire the network, wasting battery power and likely missing the broadcast channel they have woken up for. So it is expected that Weightless networks will only re-plan frequencies at most every approximately 15 minutes. Of course, if there are no changes in the network, interference levels or white space allocation then there would be no reason to re-plan. Also, algorithms that can entirely re-plan large networks are likely to take many minutes to run in any case.

So if a cell decides that it is necessary to remove a frequency from its hopping list it informs the network planning system which includes this request with any other frequency planning changes needed in that cycle.

The cell then has to decide when to reinstate the frequency. This is not an easy decision as it may not be able to determine whether the interference has disappeared unless the frequency is re-used. But then it must remain in use until the next planning cycle. There is an option in Weightless to request certain terminals to make interference measurements on given frequencies but these are not particularly reliable as the terminal may have moved or its ability to measure signal levels accurately may be poor. Solutions such as reinstating after 15 minutes, then removing again if the interference remains, then reinstating after 30 minutes, and so on, could be envisaged. At this time, the characteristics of any interference are not known so it is difficult to determine which strategies would be most appropriate. Since the entire frequency planning system runs as software in the core it can be updated at any time.

Weightless allows up to eight frequencies to be in use in any cell. This limit is in place mostly to restrict the space needed in the MAC header where the hopping sequence is described to terminals. The limit also keeps the re-planning process tractable. In practice, there are few benefits of extending the hopping sequence beyond eight frequencies and it is rare that there will be eight channels or more available and usable in TV white space bands.

At the other extreme there may be insufficient channels available to allow planning without frequent collisions of the same frequency being used in neighbouring cells. Such interference could be accommodated by using longer spreading factors or alternatively Weightless does allow for a TDMA mode. In this mode a base station only transmits for part of the possible downlink time and only receives for part of the possible uplink time. For example, if there were only a single channel available across multiple base stations then this might be split into eight parts (generally these would be equal but this is not necessary). Base station 1 might be assigned part 1 of the channel for its downlink and part 5 for its uplink.

Base station 2 might be assigned parts 2 and 6, etc. Terminals will be little affected – they will simply see fewer assignment slots. The only implication is that frame headers will no longer be synchronised across cells. It is not thought the TDMA mode will be widely used.

5.4 Code assignment

As well as assigning frequencies there are a number of other cell-specific parameters that the network needs to assign. These include code sets and scrambling sets.

In some cases the communications path makes use of direct sequence spread spectrum (DSSS) as a mechanism to extend the range at the expense of data rate. This does allow the possibility of neighbouring cells using orthogonal DS-codes such that if frequency interference occurs it may still be possible for some information to be received. The extent to which this is beneficial depends on (1) the relative power levels at the terminal (effectively the 'near–far' problem in CDMA) and (2) the timing difference between the arrival of the transmissions (and hence the extent to which orthogonality is preserved). With no special measures being taken to control either of these it is possible that in some situations the code planning brings little benefit. However, it should never actually result in worse reception than would otherwise be the case and may be beneficial and so is recommended.

For simplicity all base stations will make use of the same code sequences but with different offsets. The assignment of the offset is made such that neighbouring base stations have assignments that provide the maximum orthogonality. Code planning does not change according to frequencies assigned and so only needs to be altered when new base stations are added to the network.

For example, if the terminal is operating at a spreading factor of 63, then there are 64 available codes in total, of which each base station will use a subset of only 4 assuming a 1 in 4 code selection is being used to convey 2 bits per symbol (this is described in more detail later). This means different code subsets can be allocated to up to 16 neighbouring base stations. The network will signal to each base station which offset it is to use for each of its possible spreading codes. This will only need to be done at network initialisation and when new cells are inserted. At the time of writing, it was still being debated whether it was necessary to signal these offset values to terminals in the cell.

Each base station transmits a synchronisation word to allow terminals to acquire network timing. Neighbouring base stations should use different synchronisation words to avoid interference and false acquisition as far as possible. The same approach as adopted for code planning (above) is used to provide different synchronisation codes for neighbouring base stations, selecting from one of the eight possible codes available. At the time of writing, the exact structure of the synchronisation message was still under some debate.

5.5 Scheduling

There are a number of reasons to schedule traffic to and from terminals, including:

- To prioritise certain traffic when there is congestion.
- To normalise the received power levels from devices transmitting simultaneously on the parallel uplink channels (to be explained in more detail in later chapters).
- To ensure terminals are not addressed on frequencies where they are in antenna nulls.
- To minimise interference suffered by terminals from other white space users.
- To minimise intra-system interference.

Each of these is now described in turn.

5.5.1 Traffic prioritisation

The network will typically know in advance how much traffic there is to be sent from each base station in each frame since much of the traffic is scheduled in advance. If there is less traffic than the capacity of the frame (taking into account the need to spread transmissions to some terminals) then there is no need for complex scheduling. But if there is more traffic than capacity then the core network can choose to prioritise certain traffic, delaying or discarding the remainder. It might do this on the basis of service contracts with different customers, some of who may pay extra for higher quality of service.

5.5.2 Normalising power levels

A Weightless system will typically have 24 FDMA uplink channels for each downlink. The base station will decode these simultaneously. The dynamic range needed at the base station can be reduced if the received power level from each of the 24 transmitting terminals is similar at the base station. In conventional systems this could be achieved using power control commands from the base station. However, there is limited capability to do this in Weightless. Firstly, terminals may send short bursts with little time for power control loops to work. Secondly, terminals using spreading are all transmitting at maximum power but being received with progressively lower power for higher spreading factors. The core network can achieve somewhat similar power levels by scheduling terminals it expects to have similar received powers at the base station to transmit at the same time. Its ability to do this is limited by the predictability of the signal level from terminals which may be high for static terminals but very limited for moving terminals.

The extent to which this sort of scheduling is important will depend on the quality of the base stations actually procured by a network operator.

5.5.3 Avoiding antenna nulls

One key source of interference in white space networks can be distant TV transmitters. This type of interference is typically most problematic at the base station which has elevated antennas. It can be reduced if the base station steers a null towards the interferer using beam-forming techniques (to do this it needs to have at least two antennas at the base station). Since the TV transmitter and the base station are fixed, the direction of this null will remain constant for a particular frequency. A base station might have interference on multiple different frequencies often from multiple different TV transmitters located in different directions. So as it hops from frequency to frequency the direction of the null will vary.

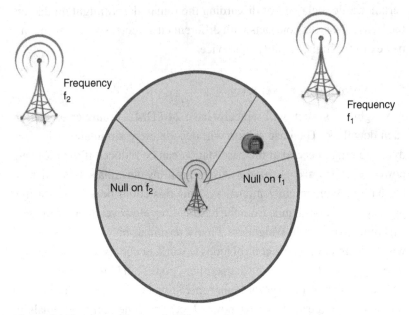

Figure 5.1 Steering nulls towards TV transmitters

A problem with antenna nulls is that any terminals in the direction of the null will receive a much attenuated signal and may not be able to communicate. It would be better to address such terminals when on a different frequency hop where the null was in a different direction. Hence, where the location of the terminal is known and beam-forming is being used in the base station it is advantageous to schedule communications with terminals when they are not in a null. This is shown in Figure 5.1 where a Weightless base station is surrounded by two TV transmitters. A smart meter is in the null steered towards the TV transmitter operating on frequency f_1 but can be communicated to when the base station is transmitting on frequency f_2 when the null is in a different direction.

5.5.4 Minimising interference

As discussed above, where there is local interference affecting certain frequencies in certain parts of the cells then terminals located in those parts can be scheduled on other frequencies where there is less interference. Again, this is only viable where the location of the terminals is approximately known.

Where there is known interference between base stations, perhaps because there are insufficient frequencies to plan without collision, scheduling can be used so that terminals in the overlap region between the cells are not addressed during the times when the frequencies clash.

5.5.5 Overall

Scheduling is not essential in Weightless – traffic could just be given resource on a first-come first-served basis. But there are substantial gains that can be achieved from clever scheduling including the ability to offer differing grades of service to clients. Taking all the factors listed above into account can make for a complex scheduling algorithm, but given that computing resources are nearly free but radio resources are expensive this is typically worthwhile. The complexity of scheduling performed is a choice for the network operator.

5.6 Calibration mode

The scheduling and frequency hopping algorithms discussed above can sometimes depend on a reasonable understanding of the noise levels on each frequency and the impact of interference from one base station to another. It can be difficult to determine this during normal operation where distinguishing signal from interference is typically not possible. Hence, it may be useful to have a 'calibration mode' where either all the base stations listen, or only one base station transmits and the others listen.

Achieving this is relatively simple. Each base station sends its header as normal then stops transmission. In the header no downlink or uplink resources are advertised as being available. Individual base stations can then be instructed to transmit by sending them pre-formed frames full of random data at a constant power level and spreading factor. Calibration can then be performed according to one or more of the following:

- When no base stations are transmitting each base station can be instructed to measure its noise floor across a number of channels.

Depending on how quickly a measurement can be made on each channel this could be performed across the network in just a few seconds.

- A selected base station can be asked to transmit a constant power carrier and all the others in the vicinity asked to make measurements on the frequency used. This can then be repeated for all base stations of interest. This could take some time but will typically only need to be performed once on initial network deployment. Assuming one measurement per frame and that clusters of ∼100 base stations are analysed the process could take 200 s (3 minutes).

5.7 Load balancing

As mentioned in the previous chapter, terminals in Weightless manage their own mobility. They autonomously decide whether to hand off to neighbouring cells. This is normally advantageous but there are some cases where network control is desirable. These include:

1. When a cell is congested but neighbouring cells are not.
2. When a new cell is inserted in the network.

In all cases, the network does not know the signal levels experienced by each terminal, so it will not know the implications of forcing it to move to a particular cell. Hence, it can only 'suggest' to the terminal that it consider moving, leaving the final decision to the terminal.

5.7.1 Load imbalance

The core network will be aware of levels of congestion on base stations throughout the network. If there is an imbalance such that there are uncongested base stations neighbouring a congested base station then the network may instigate a procedure to attempt to balance the loading. It does so by indicating to attached terminals a preference for them to reattach to these uncongested base stations via a control message containing the identities of those base stations that are uncongested. Terminals then make their own decision as to whether to hand off to a neighbouring cell based on (1) whether they have good coverage from that cell and (2) whether they are likely to generate substantial traffic levels in the coming hours. For example, it will generally make little sense for an energy meter to hand over to a new cell because the signalling load involved may be higher than messages to be transmitted during the period of congestion. It cannot be entirely left to the terminal manufacturer as to how this decision is made as this might result in unanticipated network behaviour.

The recommended algorithm that the terminal deploys is as follows:

- When a load balancing message is received estimate the predicted traffic loading to be generated in the time period indicated for congestion. If less than the signalling load associated with moving from one base station to another and back again then do nothing. This estimate can be made on the basis of the traffic levels generated during the previous time period unless there is more specific information available to the terminal radio system.
- If there is sufficient power available (e.g. it will not significantly drain the battery) measure the signal levels from other neighbouring cells and estimate the spreading factors that would need to be applied for communications with each one.
- For all target cells determine the switching propensity factor (SPF) (see below) and select the cell with the highest SPF.
- If highest SPF > 0.9 then switch randomly within 5 min. If highest SPF < 0.3 then do not switch. Otherwise switch according to time = predicted congestion time * (1-SPF).

The SPF = congestion factor × message factor / difference in spreading factor where:

Congestion factor = 1 if congestion < 50%, 0 if congestion > 90%, otherwise $1 - ((\text{congestion} - 0.5) \times 2.5)$

Message factor = 1 if message > 1500bits, 0 if message < 150bits, otherwise $(\text{message} - 150) \times (150/135)$

Where there is a base station with multiple carriers the network can perform load balancing across the carriers using a similar approach, with a handoff command to particular terminals or classes of terminals with details of the carrier to which the terminal is to attach.

5.7.2 New cell

If a new cell is inserted into a mature network, particularly where there are mostly static terminals and no new terminals are deployed, it is possible that terminals would remain attached to their existing cell even where it would be advantageous to be attached to the new cell. In this situation all terminals in a cell can be instructed to 're-scan' to check whether they are connected to the optimal base station. They do not need to do this immediately but should randomly schedule their scan and move over the following 24-hour period, avoiding times when they have scheduled traffic to transmit or receive.

5.8 The network to base station interface

The interface between the network and the base stations is standardised by Weightless. This allows base stations to be sourced from a range of manufacturers. The interface is relatively simple as the base station comprises broadly only a radio function. The network sends complete frames to the base station which converts these to RF and transmits. In return the base station demodulates received frames and converts them to a baseband signal before sending them to the network.

More specifically, the network sends to the base station all fields in the frame except the DL_SYNC which is generated locally by the base station. The data transmitted to the base station consists of a number of tagged sections, each of which correspond to a section within the downlink block. The overall structure is therefore:

- Frame identifier
- Channel
- Section list

Although the frame identifier is implicit in the DL_FCH section, it is listed explicitly in the downlink block message for convenience. The channel identifier defines the radio channel to be used for the forthcoming frame.

The section list comprises a number of entries which are provided in the same order as they must be transmitted over the air interface. Each entry contains:

- Type (DL_FCH, RS_MAP or DL_BURST)
- Spreading factor
- Length
- Contents

5.9 Summary

The Weightless core network has many of the functions of a cellular core network but is much simpler due to a range of factors relating to terminals. As a result, it is typically implemented as a virtual machine within the cloud. There is a wide range of optimisation functions performed in the core including setting the frequency hopping assignment, scheduling traffic and managing load imbalance.

6 The MAC layer

6.1 Overview

The function of the medium access layer (MAC) is to share the available radio resource among the many thousands of devices that might wish to access it. As discussed in Chapter 4, the starting point in understanding the Weightless MAC is the design options that were selected as a result of the underlying requirements and constraints. Those that concern the MAC include:

- the use of TDD
- long time duration frames
- super-frame structure with sleep mode.

There is never a clean distinction between MAC and PHY and the MAC must provide PHY-level information to the terminals such as the hopping sequence hence there will be some blurring between this and the next chapter.

6.2 Scheduled and contended access

There are two ways that a terminal can obtain resource for communications or that the network can communicate with the terminal. These are scheduled access and contended access. Scheduled access occurs when the network and the terminal have pre-arranged to communicate at a particular time. For example, when transmitting a meter reading, the network and terminal may agree the next time that a reading will be sent and this resource is reserved in advance. Contended access occurs when there is information to send that was not envisaged. For a terminal, contended access requires attempting to gain radio resource to signal a desire to communicate. For the network it requires the base station controller making space within a downlink frame to carry the traffic.

In person-based communications networks such as cellular networks, almost all communications starts with contended access. This is because the network cannot anticipate when the user might want to make or receive a call or initiate a data session. However, in a machine-based network it may be that the majority of communications is regular and hence can be anticipated. This is beneficial because when scheduling in advance it is possible to achieve close to 100% utilisation of resource. But contended access systems can rarely run at anything above 35% loading as otherwise too many collisions occur between those randomly seeking resource, leading to repeat transmissions, leading to further collisions, etc.

6.3 The frame concept

6.3.1 Introduction

In Weightless the MAC is based around the concept of frames. These are structures that repeat around every 2 s (the duration is a variable factor that can be selected by the network operator). A frame is divided into downlink resource and uplink resource. The downlink is divided into frame header information used by all terminals (that are awake) to understand the contents of the frame and then payload information destined for one or more (typically hundreds) of terminals. The unit of division on the downlink is time. The uplink is divided into contended access resource and scheduled resource. The unit of division on the uplink is a mix of frequency and time as there are typically 24 narrowband uplink channels available which in turn are divided into time. However, this is more of a PHY issue and will be returned to in the next chapter.

Frames are grouped together into super-frames. A super-frame can contain any number of frames but typically will be 512. With a 2 s frame duration this gives a super-frame duration of just over 17 minutes. As explained earlier a super-frame duration of the order of 15 minutes or so is considered optimal for Weightless. The first six frames of a super-frame have an identical structure to all other frames but terminals are encouraged to listen to these as it is where the network will place

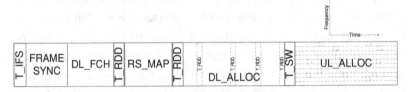

Figure 6.1 Frame diagram [Source: Weightless *standard*]

control information and unscheduled paging information. These frames are known as 'broadcast frames' because the information they contain is intended to be broadcast to all terminals in the cell.

A more detailed representation of a frame is shown in Figure 6.1.

The frame starts with the inter-frame separation time (T_IFS) to allow for local oscillator retuning and set-up at the base station. It is set at 250 μs, timed from the end of the last decoded bit received from the previous frame.

Next is a downlink synchronisation (FRAME_SYNC) sequence that is used by terminals to establish sufficiently good frequency and timing estimates that subsequent sections are only prefixed with a burst (short) sync sequence. The synchronisation sequences used are discussed in more detail in the next chapter.

The FRAME_SYNC is immediately followed by the downlink frequency and information channel (DL_FCH). This provides all the information needed by a terminal to understand the parameters in use in the cell including the cell identity, the frequency hopping sequence and the relative position within the super-frame. This means that a terminal that has woken up or tuned into the cell for the first time is able to decode all the information it needs from this short burst, minimising battery requirements. Indeed, each frame could be considered 'stateless' in that the contents and organisation of a frame does not depend on any other frame transmitted.

The DL_FCH must be transmitted using the worst-case modulation and coding scheme in use in the network. This will typically mean BPSK modulation and a spreading factor of 512 or more (see Chapter 7). The key reason for this is so that terminals on the edge of the cell are able to decode the DL_FCH. On this basis it might be assumed that the coding scheme could be the worst in the cell rather than the network, however, this would be problematic for terminals trying to acquire the network which would not know what modulation and coding to use in their initial decoding of the frame (acquisition is discussed in more detail in Section 7.13). The net result is that the DL_FCH can take a relatively long time. The DL_FCH contains details of the coding scheme used in the next field in the frame, allowing more flexibility for subsequent parts of the frame. The contents of the DL_FCH are described in more detail below.

After the DL_FCH is a gap of at least the time to receive data decode (T_RDD). This is because there is a delay required at the receiver to decode and process any section after the bits have been sent over the air which depends on the spreading factor used. To enable this T_RDD needs to be at least 200 μs plus twice the symbol period of the modulation scheme being used.

Next is the resource map (RS_MAP). This describes how the downlink and uplink resource is to be shared for the entire frame. It is split between downlink and uplink. On the downlink, each element of resource (this will be described in more detail later) is allocated to a particular identity which may relate to an individual terminal or a group. The uplink resource is split between scheduled and contended resource. The scheduled resource is allocated to particular terminals, while the contended resource is available to groups of terminals (one group may be 'all terminals') to use for unscheduled transmissions. The way that this is encoded is the same regardless of downlink, uplink, scheduled or contended and consists of the identity, number of slots allocated and modulation/coding scheme in use. The resource map also needs to be heard by all terminals in the cell and so is typically transmitted using a high spreading factor. However, unlike the DL_FCH it does not need to be the same spreading factor for all cells since the spreading factor is specified in the DL_FCH. It can therefore be at a spreading factor appropriate for the cell size – so small cells might use lower spreading factors on their RS_MAP. The structure of the RS_MAP is described in more detail below.

After another T_RDD period to allow terminals time to decode the RS_MAP before needing to decode data, the downlink allocated channels (DL_ALLOC) is transmitted. This contains the actual data for terminals or groups of terminals. The structure of a DL_ALLOC burst is described below.

A switching time of 150 μs is allowed for terminals and base stations to move from one TDD mode to the other (base stations from transmit to receive, vice versa for terminals). Then the uplink channel allocation (UL_ALLOC) follows in a similar manner to the downlink. The structure of the uplink allocation is described in more detail below.

6.3.2 The FCH

The function of the DL_FCH is to provide a terminal with all the information it needs to decide whether to (1) return to sleep mode until, e.g., a broadcast frame or (2) continue to listen to decode the resource map.

It starts with a 32-bit synchronisation word which is used for finer timing synchronisation than the 'frame' synchronisation burst – essentially there is a two-stage process of coarse and then fine timing as described in more detail in Chapter 7. The data is then provided as follows:

- BS_ID. The base station identifier (24-bits) enabling terminals to check that they are listening to the base station they expected and also to determine whether the base station is part of a network which they have rights to access. To aid this the first few bits of the base station address form a network code.
- FRAME_COUNT. The MAC frame count (16-bits) giving terminals a point of timing reference. This increments with every frame and allows terminals to determine how long it will be to the next broadcast channel and when given future resource in terms of frame number, how many frames will occur before this resource is available.
- HOP_CH_MAP. Hopping sequence channel map (48-bits) – indicating which channels the base station is using in its hopping sequence. This consists of a list of up to eight 6-bit numbers corresponding to the channels to be used in the hopping sequence. If the hopping sequence is shorter than eight channels then the channel map value is set to zero for unused channels. Note that other, more efficient, schemes were considered to encode this information such as bitmaps and differentially encoded numbering but it was thought this might

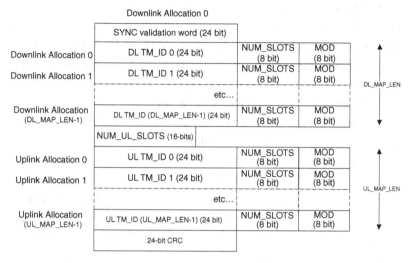

Figure 6.2 RS_MAP schematic [Source: Weightless *standard*]

reduce flexibility in future cases where different frequency allocations or channel bandwidths were in use.

- DL_MAP_LEN. The number of allocations in the downlink resource
  map (RS_MAP) (8-bits). If this is zero, the frame will contain no
  downlink resource map, i.e. it will be uplink-only or an empty frame.
- UL_MAP_LEN. The number of allocations in the uplink resource
  map (RS_MAP) (8-bits). If this is zero, the frame will contain no
  uplink resource map, i.e. it will be downlink-only or an empty frame.
- RS_MAP_MOD. (8-bits). This details the modulation scheme and
  coding rate that the RS_MAP will be transmitted at.
- FRAME_TIME (2-bits). The length of the frame, allowing a small
  variation (e.g. 1 s, 2 s).
- Reserved bits for future use (14-bits). These are provided under the
  assumption that future developments of the specification may require
  additional information in the FCH.

6.3.3 The RS_MAP

The structure of the RS_MAP is shown in Figure 6.2. Like all other fields
the RS_MAP starts with a short synchronisation burst to achieve precise

timing (even though the terminal will have achieved precise timing at the start of the FCH it may have drifted by the time the RS_MAP is broadcast and hence it is appropriate to re-synchronise). There then follows a 24-bit sync validation word which is a fixed pattern of bits and allows terminals to determine whether they are correctly synchronised and receiving the base station. If they do not receive the expected bit pattern in the validation word then they do not attempt any further decoding but instead return to frame-level synchronisation, possibly on the next frame.

The bulk of the RS_MAP is made up of a table of terminal identifiers, the number of slots that they have been allocated and the modulation scheme used. The RS_MAP ends with a 24-bit cyclic redundancy check (CRC) which allows the terminal to verify that it has received the RS_MAP correctly. If it has not it discards the information and takes no further action for the frame.

The terminal identifiers are not necessarily in numerical order within each part of the MAP allowing optimal packing on the uplink where, for example, terminals with similar received power levels might be scheduled simultaneously on the multiple FDMA channels, as discussed in the previous chapter. On the downlink there does not currently appear to be any need for non-numeric ordering but this facility is provided in case such a reason should emerge in the future. In any case, terminals need to listen to the entire RS_MAP in order to be able to use the CRC so there is no advantage in restricting the ordering to be numeric.

For the downlink allocations, the table is ordered in the order in time that the slots appear in the DL_ALLOC part of the frame. A terminal can therefore work out where its allocated slots start by summing the preceding NUM_SLOTS allocations in the table.

In the uplink table, the available resource allocations span both sub-channels (frequency) and time slots. A fixed uplink time for the frame is defined for all sub-channels by NUM_UL_SLOTS; this is chosen by the base station to be no longer than the amount of time available in the frame after DL_FCH, RS_MAP and DL_ALLOC sections have been completed, allowing for T_IFS, T_SW and T_RDD where appropriate. The uplink resource is numbered such that slots [0 ... NUM_UL_SLOTS) indicate sub-channel 0, [NUM_UL_SLOTS ... 2*NUM_UL_SLOTS) indicate sub-channel 1, etc. The uplink resource is then allocated in the same way as the downlink resource, i.e. allocation 0 allocates NUM_SLOTS time on sub-channel 0, allocation 1 allocates any remaining time in sub-channel 0 up to NUM_UL_SLOTS, etc.

In the uplink table, the base station will never allocate resource such that they span multiple sub-channels, and will only utilize at most one sub-channel per addressed terminal. This makes terminal design simpler.

The time slots are quantised in fixed units of 1 ms for the downlink section, and 10 ms for the uplink section. The reason for the longer time on the uplink is that the data rates are lower on each of the narrowband uplink channels than on the downlink. Quantisation is helpful to ensure a known start time and to be able to quantify the amount of resource allocated using a relatively small number of bits. It may result in some slight inefficiencies where the physical resource available is not an exact multiple of the slot duration.

6.3.4 The DL_ALLOC and UL_ALLOC

The DL_ALLOC section of the frame is where the main payload data transfer between base station and terminal takes place. It is divided into separate slot allocations to individual or broadcast terminals, as defined in the RS_MAP for the frame. Each resource allocation is no longer than the number of allocated slots less T_RDD and is known as a 'burst'.

Each burst consists of a PHY burst sync sequence, a 24-bit sync validation word, and modulated data consisting of one or more burst payload data (BPD) structures. The PHY burst (short) sync sequence allows the receiver to obtain fine frequency error and channel estimate (using the information from the long frame sync for coarse information). This is followed by a modulated 24-bit sync validation word (fixed 0xFAC688) used to validate the sync and indicate the start of the first bit of data. The structure of a burst within an allocation in DL_ALLOC is shown in Figure 6.3.

Each BPD structure consists of a header, payload and CRCs. There are a number of different types of BPD structure described in a subsequent

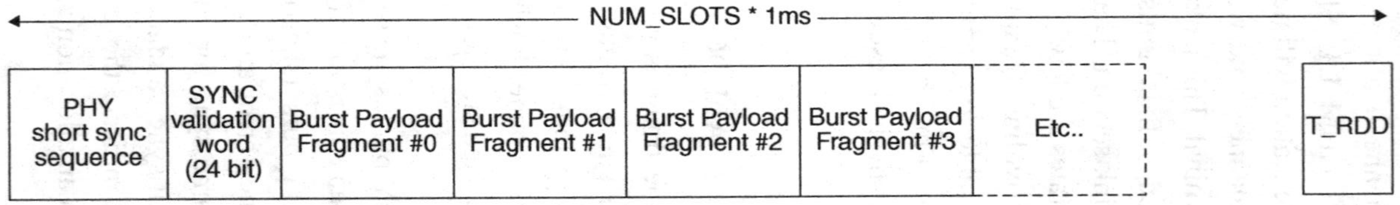

Figure 6.3 Burst structure [Source: Weightless *standard*]

section. The last BPD in any burst has a flag indicating that no more BPDs follow.

The UL_ALLOC broadly follows the same format as the DL_ALLOC channel.

6.4 Acknowledgement

Acknowledgement that a message has been received correctly is often an important part of machine communications. It would have been possible to design Weightless networks without their own acknowledgement, relying on higher-level acknowledgement, for example as is provided within the TCP/IP control stack. However, this is typically inefficient as it operates over time frames longer than that which would be optimal for a wireless system. Hence, Weightless has its own acknowledgement system although it is expected that higher-level systems may still be used as well (if Weightless functions correctly these will add little value, but equally cause no problems).

The Weightless acknowledgement system is based around the concept of sequence numbers. Each block of data to be transmitted is given a sequence number (SN) where a block can have any length between 0 and 65 535 octets. The division of data into blocks takes place outside of the Weightless protocol at the application level (although applications do need to be sure that they do not exceed the maximum length). Sequence numbers are assigned sequentially with separate numbering for the downlink and the uplink up to the maximum for the 8-bit field (256) when they reset to 0.

When a block is transmitted its SN is included within the block header. When the receiver sends a message back it includes the next expected SN (NESN) allowing the transmitter to determine whether the previous block needs re-transmission. If there is a significant delay between block transmission and acknowledgement and additional blocks to transmit then the transmitter can continue to send blocks without acknowledgement up to a system-defined maximum. The re-transmission behaviour is up to the network or the terminal. For example, if blocks 1–5 were sent and the response had a NESN of block 2, the transmitter could at one extreme send block 2 only and wait to see whether the NESN after this was 6. Alternatively, it could re-transmit blocks 2–5 in any case. Or it could select strategies between this. Because each BPD has a SN it is possible for the receiver to determine what has been selected. However, profligate re-transmission does result in increased use of network resources and can be expected to generate higher costs for the end user.

As well as effectively controlling retransmission via the NESN, a receiver can request a retransmission of part of all of a BPD. To do this a Retransmit-SN message is sent to request a whole BPD or a Retransmit-Partial-SN can be used where the start and end offset within the BPD are included. These are control messages and as such are sent as the payload of a BPD. However, the SN and NESN are set to 0 to indicate this is not part of a data sequence.

6.5 Mapping data to resource allocations

The Weightless MAC needs to take data destined for terminals or from terminals in a format suitable for the application and process this data so it can be accommodated within the physical layer transmission resource available within Weightless. Applications will deliver data in blocks, partitioned in a manner which makes sense for the application and which could be almost any size. At the other extreme, the physical layer, as discussed in the next chapter, has strictly defined bit periods and capacity linked to issues such as the propagation environment and number of terminals in the cell.

Broadly, the application delivers blocks of data in its preferred size and the physical layer and MAC determine the burst size that can be assigned to a terminal based on the radio resources available. Blocks then need to be matched to bursts. There are then two issues:

- If the blocks are bigger than the bursts then they need to be segmented across multiple bursts (they will never be smaller as the burst is sized according to the need).
- If the bursts are long it may be sensible to fragment the block with each fragment having its own CRC otherwise the probability of error in the burst can become overly large requiring constant

re-transmission. Fragmenting the block can mean that only some fragments are received in error and hence only these need to be re-transmitted. The optimal fragment size depends entirely on the typical error rate on the link. In Weightless this is set at around 5%–10% fragment error rate as discussed in detail in Chapter 7.

This leads to four different BPD types:

1. Containing an entire block which is not fragmented.
2. Containing an entire block, fragmented into sub-parts.
3. Containing part of a block, which is not fragmented (and hence there will be further BPDs with the remainder of the block).
4. Containing part of a block which is fragmented.

The exact details of the field structure for each of these different BPD types are defined in the *standard*. An example of the second in the list above is provided in Figure 6.4.

This consists of flags, length (0–65 535 octets), sequence numbers, fragment length, plus 24-bit CRC over the header, plus payload of length 0–65 535 bytes split into fragments of length 'FragLen', each with a 24-bit CRC calculated over the fragment. The final fragment will have a length corresponding to the remainder of the data (Length modulus FragLen).

There can also be multiple BPDs within a burst, which could be of different types. So, for example, if the application delivers two short blocks, these could be encapsulated in separate BPDs which are then sent in the same burst.

6.6 Terminal and group identities

As in any cellular-like wireless system, a unique and well-defined system of identifying terminals is important. It is quite possible that clients will have terminal identities that may be based on their own numbering system or an IPv6 address but these cannot be relied upon within Weightless to be unique and of a consistent format. Hence, Weightless assigns a unique identity to each terminal. This can be mapped onto a client identity or

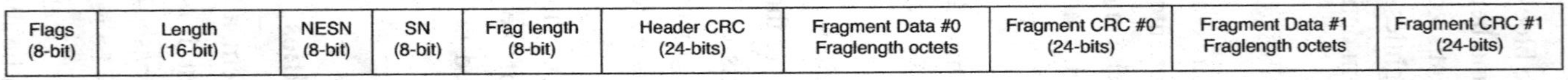

Flags (8-bit)	Length (16-bit)	NESN (8-bit)	SN (8-bit)	Frag length (8-bit)	Header CRC (24-bits)	Fragment Data #0 Fraglength octets	Fragment CRC #0 (24-bits)	Fragment Data #1 Fraglength octets	Fragment CRC #1 (24-bits)

Figure 6.4 BPD structure (example) [Source: Weightless *standard*]

IPv6 address within the core network so that the terminal is addressed from the client using their preferred addressing scheme.

With a potentially large number of terminals worldwide – perhaps well in excess of 100 billion, a long identity is needed so that there is no address shortage. Hence terminals are assigned a 128 bit unique user terminal identification (UUTID) at the time they are manufactured or deployed. This might be embedded or programmed onto the chip and is unchanged for the lifetime of the device. It corresponds approximately to the international mobile equipment identity (IMEI) on a cell phone.

Using such a large identifier during normal operation would be inefficient. With 30 byte messages, an extra 16 bytes would need to be added to the message flow simply to specify the terminal. Hence, when a terminal attaches to a base station, it is assigned a temporary identity (TID) which is used to identify it for subsequent communications with that base station. The size of this identity only needs to be large enough for the expected maximum number of terminals per base station, rather than worldwide, so it can be much smaller. In Weightless TIDs are 24 bits long. The TID is 'temporary' in that it can be reassigned to another terminal. However, for some terminals that are static, such as smart meters, and which never change cell, the TID might be assigned once at the time the terminal registers and remain unchanged for the lifetime of the terminal.

The TID is assigned by the network when the terminal first attaches to the cell. The terminal sends a message with its UUTID and in response is assigned a TID. It needs to store this TID as it will be used in all subsequent communications. Should it 'forget' the TID, perhaps due to a power failure or re-boot, it re-starts the acquisition. The network will note that the UUTID is already assigned a TID and 'remind' the terminal of its TID.

There are two classes of TID – those related to individual devices known as individual TIDs (iTIDs) and those related to groups, known as group TIDs (gTIDs). Both are 24 bits long and drawn from the same numbering range. However, gTIDs apply across the entire network whereas iTIDs only apply to an individual cell. Hence, the numbering range for gTIDs must be reserved and not used by any cells within the network. This is simply achieved in Weightless by the first bit of a TID being '1' for a gTID and '0' for an iTID.

A terminal can belong to any number of groups. For example, a smart meter might belong to the following:

- all smart meters
- smart meters deployed in the UK
- smart meters made by Itron
- smart meters connected to substation 2345
- smart meters with home control functionality
- etc.

Terminals can be pre-populated with group identities or they can be sent to the terminal by the network using a control message. A terminal can request an update of the groups it belongs to at any point. Further control messages are available to remove a terminal from a group and generally manage the group identities. One complexity is for group messages related to area. If a terminal moves out of the area it should be removed from such a group (and likely added to a new one). This is handled by the network which detects the attach message when the terminal registers in a new cell and processes this to determine whether any group identity changes are required, sending update messages to the terminal and the network group membership register as needed.

The similarities between iTIDs and gTIDs make broadcast calls (calls destined for multiple terminals) very simple from a protocol viewpoint. The network simply sends the message to a gTID and all terminals in this group receive it. There are some details around which cells to transmit the message in and the mechanisms for acknowledgement that are discussed further in Section 6.9.

Group identities can also be used as part of the contended access process by indicating that certain contended access slots are reserved for particular groups of terminals. One of these is the group of 'all terminals' and a particular gTID is reserved for this for which all terminals are permanently members. More information on the contended access process is provided in Section 6.7.

In normal operation a terminal will either continue to communicate periodically with a base station using its iTID, will move to a different base station and reattach, or will explicitly detach from a base station. Any of these behaviours will allow the base station to determine the iTID is no longer needed and return it to the pool of available iTIDs. However, there are situations when the terminal will cease activity without informing the base station, for example because it is destroyed, its battery is removed or similar. This could result in iTIDs being unnecessarily reserved. To prevent this, the base station will de-allocate an iTID if it has not received any communication from a terminal for a pre-set period such as 30 days. Terminals must retain information on their last transmission. If a terminal determines it has not transmitted for more than the de-allocation period it must re-attach before transmission. If a terminal does not have a clock providing elapsed time it must determine time from other sources. One might be a time control message broadcast in the super-frame. If a terminal has lost all record of its previous transmission it must assume its iTID has expired and perform a reattach.

Note that unlike cellular systems, terminals are not required to have 'SIM cards'. Indeed, the 'S' in SIM means 'subscriber' and there is typically no subscriber associated with a terminal. One of the key functions of a SIM is to enable a subscriber to keep the same phone number when changing phone, but this is not relevant to terminals which will likely never change wireless module. Hence, SIM cards are not mandated for Weightless. However, if a terminal manufacturer wishes to have a removable element in their terminal design which contains various identities they are free to do so – but this is not standardised across all terminals.

In cellular systems the SIM also handles security. This is because the necessary subscriber-related keys need to be stored on the SIM so that when changing mobile these keys are still available. This is not relevant to Weightless where all security functions can be handled within the main chipset. Hence, Weightless has cellular-grade security but no SIM.

6.7 The contended access process

6.7.1 Introduction

Terminals will not in general use contended access but will await resource that they have been pre-assigned. However, there are two cases where contended access (CA) may be needed:

- to attach to the base station
- when the terminal has an unscheduled message to transmit.

Contended access is relatively complex because of the uncertainty as to when terminals might seek resource. Further, contended access systems have the possibility of failure if the contended access load rises above around 35% of the channel capacity. In this case, there will be a high chance of collisions of CA requests, resulting in re-transmission. The retransmission raises the loading on the channel further increasing the chance of collision until a situation is reached where there is no through-put because there are collisions on every burst. In cellular systems this outcome can occur when there is an event, such as an emergency, in an area which causes many subscribers to simultaneously attempt to make a call. In machine networks there are similar situations, for example when a power failure causes all meters in the region to send an alert message. Hence, it is essential that Weightless has mechanisms to prevent some of the terminals accessing the network, reducing the overall load.

The tools Weightless has at its disposal include:

- Changing the number of CA slots in any frame (for example increasing CA resource at the expense of uplink or downlink).
- Restricting access to CA slots by using a gTID other than 'all terminals' so that certain classes of terminals have a higher priority access.
- Making the CA slots shorter. This results in more CA slots, but means less information is transmitted on each slot. This is discussed in more detail below.
- Forestalling multiple alerts. This is achieved by the network realising that it is receiving many messages from the same event (e.g. a local power failure) and pre-emptively sending a message to all terminals of a particular group (e.g. smart meters) telling them to cancel any 'alert' message they may be about to send. It requires that terminals are able to distinguish between alert messages and other messages and set a flag internally to know whether they are in 'alert state' or not.

It is up to the network which of these tools it wishes to deploy in any given situation. Clearly the forestalling of multiple alerts is useful where CA load is event-driven. Where loading is just generally high then making available more slots through increased allocation of resource to CA and shorter CA slot length may be appropriate.

Length of CA slot

The length of the slot is a trade-off. A long slot might allow a terminal to send its entire message as part of its CA access attempt with the result that it is not necessary to subsequently assign any dedicated resource to it. However, long messages will reduce the number of CA slots available. Shorter messages will typically result in the need to subsequently assign resource but have a higher probability of getting through. Very short messages might prevent even the terminal identity being transmitted in full requiring quite extended signalling flows as set out below. The network can set the length of CA slot differently for each base station and for each frame. The slot length is advertised in the RS_MAP for the frame along with the location of the CA resource, the spreading factor to use and the gTID of the terminals allowed to access it.

The CA process is now explained in more detail.

6.7.2 Contended access: initial message flow

All contended access starts as follows. On decoding the RS_MAP field within a downlink frame the terminal will be able to determine the number of CA resource bursts available to groups it belongs to in that frame and their size. It may be that the CA slots have differing spreading factors. In this case the terminal should choose the set of slots with what it believes is the lowest spreading factor that will allow it to send its message. It can base its estimate on the signal level with which it has received the downlink. In addition, if it has previously been in communication with the base station and believes it has not moved since (e.g. because it knows it is fixed) it can use the spreading factor of its last communications if this is less than the spreading factor based on base station signal level.

It then selects a random number between 0 and the number of CA resource bursts at this spreading factor and attempts to access this CA resource.

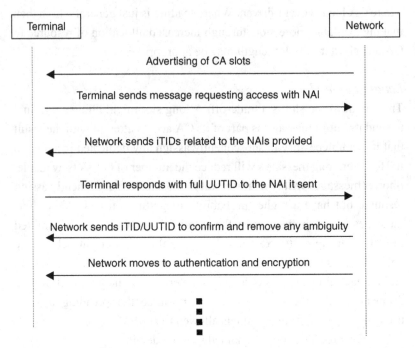

Figure 6.5 Contended access process [Source: Adapted from Weightless *standard*]

The process for the two CA cases differs from this point since in the first case there is no temporary ID (TID) assigned to the terminal.

6.7.3 Contended access for attach

As described above, the form of the contended access will depend in part on the size of CA resource bursts offered by the base station. A diagram of the message flow for contended access for attach is shown in Figure 6.5.

If the burst length is shorter than its UUTID then the terminal sends a random number of length equal to the burst length (which it stores for reference) termed the network access indicator (NAI). This may result in multiple terminals requesting resource using the same random number and hence procedures, detailed below, are needed to resolve this.

The network next provides individual temporary IDs (iTIDs) for each CA request that was for access that it received in that frame. It does so by telling terminals where these messages will be sent with a message on the RS_MAP using a reserved terminal identity (which has meaning 'all terminals that made a CA attempt for access in the previous frame'). It sends as many of these resource reservation messages as there were access attempts. It then follows this with a set of downlink messages in the assigned channels linking each of the NAIs that it has received with the iTIDs it has assigned. The terminals now know the iTID that has been assigned to them.

Note at this stage there may be multiple terminals that think they have been assigned the same iTID if they happened to use the same NAI.

If the terminal had initially provided its UUTID then the base station can move directly to the link establishment processes such as encryption. If it provided an NAI then it needs to provide its UUTID before the link can be fully established. To do this, the base station assigns uplink resource for the iTID to allow the terminal to provide its UUTID. The terminal responds with a control message providing UUTID. The network then sends a downlink control message to the iTID with the linkage between TID and UUTID. At this point, if there were multiple terminals with the same iTID only one would see a correct link of iTID to UUTID. The other terminals delete the iTID from their internal memory and start the access attempt afresh.

6.7.4 Contended access for resource

In the case of making a contended access request to obtain uplink resource the terminal already has an iTID and wishes to send a message to the network. Prior to instigating a CA message it selects a random back-off of up to n frames. This random back-off is to prevent multiple terminals stimulated by the same external event (e.g. a mains power failure) from attempting contended access in the same frame with the resulting high probability of collision. Random back-off is also applied for attach but with a smaller random value as it is not expected attach messages will be stimulated by external events but there is some possibility this might occur.

The value of n to use for random back-off can be set by the network. Terminals initially use a value of 5 but the network can change this using a control message. This control message would normally be transmitted in the super-frame. However, terminals intending contended access for resource should monitor each frame prior to the one they have randomly selected. If they receive a control message increasing n during this period they should recalculate the random back-off and re-start the counter. This allows congestion on the CA channels to be reduced. If there is a control message reducing n then they do not change their counter.

Having waited for the back-off period, the terminal selects a CA slot as before. If there is enough available space the terminal sends its iTID, otherwise a random number is used.

The process that follows then depends on whether an iTID is provided.

Case (1): iTID provided.

In this case the network can move directly to either provide resource or acknowledge the transmission. In the case that no data is provided in the CA message, or the message indicates there is more data to follow, the network provides uplink resource against the iTID to allow for data transmission. It might be possible to determine the size of the resource provision from the CA message using the data length field and the amount of data provided in the initial message. If this has not been provided, it is for the network to decide on the amount of resource to allocate. It may have prior knowledge of the size of the message that this terminal typically wishes to send, or it may provide resource depending on the free space available in the forthcoming frame. In the case that the resource provided by the network is insufficient for the terminal to send its complete message, the terminal can indicate within its uplink message the size of the resource it needs to send the remainder of its message. It does this by setting the 'more data' flag in the header in which case the last 8 bits of the message indicate the size of the additional resource needed in octets.

Case (2): NAI provided.

In this case, the network follows a process similar to the 'CA for access' process defined above. The process is shown in Figure 6.6.

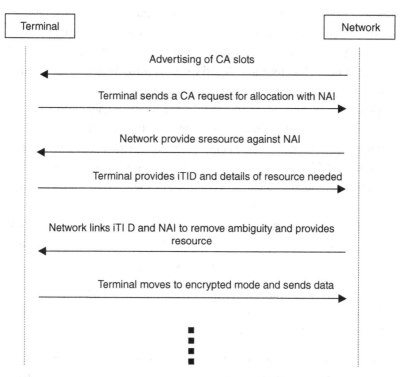

Figure 6.6 Contended access process (part 2) [Source: Adapted from Weightless *standard*]

The main difference here is that no iTID needs to be assigned during this process and so the terminal can move directly to encrypted transmission since the authentication step has already been performed (during the initial access). Normal message flow with acknowledgement, etc., can then continue as needed.

6.8 Control messages

As well as transmitting data to and from terminals, there is a wide range of control messages needed within the network to instruct terminals to undertake particular action, or from the terminals typically in response to messages from the base station. A typical example is a control message informing terminals that the hopping sequence is due to change. This message would contain the new hopping sequence and the frame number at which the change will take place. Other messages include a load balancing request, a re-scan request, updates to group membership and requests to establish location. A full list of control messages is provided in the *standard*.

Most control messages are destined for all terminals and hence are transmitted in the broadcast channel. Indeed, the typical content of the broadcast channel will only be control messages. However, some, such as a location establishment request, might be sent to a specific terminal. The network might include such a control message within the next scheduled communication with the terminal.

Control messages are layered on top of the BPD structure and are identified by the value of the Control/Data bit in the BPD header. Control messages have a standard header comprising:

- Type. Used to indicate the type of control message (e.g., 'Detach_req').
- Length. The length of the payload section.

A brief explanation of each of the control messages is provided below.

Channel Quality Request: This message may be used by the base station to obtain a measure of channel quality as observed by the terminal. The response may contain information such as signal level, bit error rate, CRC error rate and synchronisation failure rate (based on whether the validation word can be read after the synch sequence). This can be useful in optimising connection parameters for that terminal such as the modulation scheme being used. The base station can ask the terminal to measure signal levels on alternative channels in order to determine whether to reinstate a channel on which there was previously interference. However, quality reports cannot be relied upon as the terminal may be in a channel fade or may not be accurately calibrated.

Rendezvous Assignment Request: This message is sent from the base station in order to inform the terminal of future points where it will either be expected to transmit or receive data. This allows the terminal to use low-power modes between these points. The terminal may then wake up at these points and monitor the frame for downlink transmissions or uplink opportunities. The request may specify a window within which the resource will be granted rather than a single precise point, giving the network more flexibility but increasing the power requirements for the terminals. Once a rendezvous point is set the base station can schedule additional points prior to the set point using the Additional Rendezvous Request message.

Super-frame Parameter Request: This message is sent from the base station to all terminals to inform terminals of the super-frame interval and starting point. It can also include a length if the broadcast frame extends over more than a frame. It is expected to be included in all super-frames but may also be transmitted in other frames. No response is expected from terminals.

Frequency Change Request: This message is sent from the base station to all terminals to inform them that the frequencies used by the base station will change. The message will typically be sent during a super-frame but may be repeated in other frames. No response is expected from terminals. The message includes the new set of hopping channels and the frame number at which the change will take place.

Load Balancing Request: This message is sent from the base station to all terminals informing them of congestion conditions in their cell and in neighbouring cells and encouraging them to instigate a load balancing process where they are able to. No direct response is expected from terminals but as discussed in Section 5.7 the terminals may choose to change to a different cell as a result. The message contains a list of neighbouring cells that might be considered including their base station identifier, their key information including hopping sequence and their congestion levels. The message also includes a time when the next update message will be issued.

Rescan Request: This message is sent from the base station to terminals to encourage them to instigate a rescan for the best base station where they are able to. It is typically sent in the super-frame when a new cell has been inserted in the vicinity but may also be sent to individual terminals if the network determines that an alternate base station may provide a higher quality of connection than the current base station. No direct response is expected from terminals but they will re-scan and may change to a different cell as a result. The message can contain information on surrounding base stations including their hopping list to allow terminals to quickly synchronise with them.

Disallowed Base Station Request: This message is sent from a base station to individual terminals or groups of terminals informing them of base stations they are not allowed to attach to. These will typically be private networks which do not allow access of terminals outside of the private user group.

Group Membership Update Request: This message is sent from the base station to a terminal or group of terminals updating the groups to which they belong and contains a list of group identities along with whether they are to be added or removed from the list.

Group Membership Query Request: This message may be sent from a terminal in order to re-establish which groups it is currently a member of.

Establish Location Request: This message is sent from the base station to request that the terminal collects information to establish its location. The first field informs the terminal of which mode of location updating it is to use (these are explained in Section 8.3). A list of neighbouring base stations may then be provided including the base station identifier and base station map (channel list, hopping sequence and frame start offset). A duration field sets the maximum time for which the terminal should attempt the procedure. The base station will allocate an uplink opportunity for the terminal response after this time. The terminal responds with a message including the mode it adopted for location measurement and depending on the mode, a list of measurement results (for some modes the measurements are performed in the network and the terminal merely indicates it has completed the process).

Power Control Request: This message is sent from the base station to a terminal or group of terminals to instruct them to limit their maximum power. This can be used to conform to regulation for a particular white space channel or to balance received power levels across multiple terminals at the base station.

Parameter Update Request: This message is sent from the base station to vary a range of parameters used in Weightless algorithms. It contains details of the parameter to be changed and its new value.

Detach Request: This message is sent from a terminal informing the network it is detaching from the base station. It has no parameters.

Location Update Request: This message is sent from a terminal to inform the network that the terminal may have changed location. The network may react by issuing a subsequent location establishment request.

It is expected that further control messages will be added as the *standard* matures.

6.9 Broadcast messages

In discussing terminal identities it has already been noted that a terminal can belong to a group. There are many possible reasons for broadcasting a message to a group. These include a widespread software download, a request for multiple devices to take action or a daily newspaper delivery to a number of tablets. It is well understood that it is more efficient to send this message once in each cell where there is a registered terminal rather than separately to each terminal, especially where there are multiple terminals in a cell that are members of the group.

When the network has a broadcast message to send it consults its group membership database to determine the identity (UUTID) of those to send it to. The group membership database is updated by clients, for example a newspaper company might regularly change the list of those registered to receive the daily download. The client updates the list using commands from their IT system via their bespoke interface into the Weightless core network software. The network then consults the location database to determine the cells within which these terminals reside and builds a list of cells for which the broadcast message is destined. It can then build the broadcast message into an appropriate frame for transmission to the base station. Broadcast messages do not necessarily need to be sent in all impacted cells at the same time.

Broadcast messages can be scheduled or unscheduled. Scheduled messages are those where the timeslot used to broadcast the next message has been transmitted during the previous one. These are expected to be used rarely since they may not be received by a new joiner to the broadcast group or a terminal that roams from one cell to another. Unscheduled messages are those where the terminal does not have definite knowledge of when the message will appear. Unscheduled messages are advertised during the super-frame using the group address. Terminals receive this message in the same manner they would a message sent direct to them. Note that the broadcast message will need to be sent with the highest spreading factor relevant to terminals registered to the group in the cell which may be known from previous communications. If not, it should be sent at the highest spreading factor in use in that cell.

Some broadcast messages might require acknowledgement. This could be achieved using higher-level protocols outside of the Weightless system or via the procedures described below according to the preference of the client and network. If acknowledgement is needed it will be defined in the message itself using a control message appended to the end of the broadcast data. The message will define whether acknowledgement is to be provided according to one of four modes:

1. No acknowledgement needed. Terminals take no further action.
2. CA acknowledgement. Terminals send CA messages to acknowledge receipt using the standard CA process.
3. Prior reservation acknowledgement. Terminals use their next reserved slot (which has been previously agreed between terminal and network) to send any acknowledgement. This may be a relatively slow process since reserved slots may be some hours or days in the future.
4. Defined reservation acknowledgement. The network will provide uplink slots for each terminal registered in the cell for them to provide acknowledgement. In this case a subsequent control message sets out the number of frames over which these acknowledgements will be spread (and hence the terminals should monitor the RS_MAP).

Which of these the network selects will depend on factors such as:

- the urgency of acknowledgement
- the number of terminals in the cell
- the current loading on the cell.

In the case that some of the terminals did not receive the message then the network decides whether to either enter into one-to-one conversation with those terminals or to rebroadcast all or parts of the message within the cell. The network will make this selection according to the number of terminals that did not receive the message correctly.

In the case where the network decides to rebroadcast the message it initiates another message broadcast process to the same group. It sets the SN to the same value which was used originally to broadcast the message. Those group members that previously successfully received the message do not need to listen any further. Other group members listen to the message and acknowledge as required.

This process can then repeat as needed.

7   The physical layer

7.1 Introduction

The main purpose of the physical layer (PHY) is to take binary data as formatted into frames by the MAC layer and to reformat the data into a form suitable for transmission over the radio interface. The radio interface has to overcome problematic propagation, interference, multipath interference, varying signals and more and many techniques such as frequency hopping are employed. This chapter describes the physical layer in use by Weightless.

7.2 Overview of the PHY layer

The PHY layer consists of a number of manipulations performed on the MAC-level signal as shown in Figure 7.1.

Exactly how these manipulations are performed depends in part on whether transmission is downlink (base station to terminal) or uplink (terminal to base station) and the regulatory environment. In this chapter, each of the blocks shown in Figure 7.1 will be described in general and then separate subsections will look at the particular implementations for different circumstances.

In overview, the functions of each block are:

- *Forward error correction (FEC) encoding*. This adds extra redundant bits to the MAC level message in order that errors can be corrected. The amount of extra information is selected to correct the likely level of errors while minimising the overhead required.
- *Whitening*. This randomises the bit stream by multiplying it by a known random sequence to make it approximate to white noise. This overcomes problems that can be caused if the data contains long strings of 1's or 0's which might confuse synchronisation systems or result in unwanted spurious emissions.

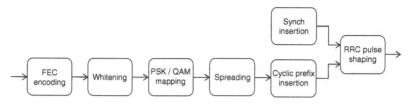

Figure 7.1 Overview of the PHY layer [Source: Weightless *standard*]

- *Phase shift keying (PSK) or quadrature amplitude modulation (QAM) mapping.* This encodes the data onto 'symbols' representing complex points in a transmitted constellation (corresponding to the phase and amplitude of the transmitted waveform). The encoding used depends on the signal-to-noise level available on the link.
- *Spreading.* This multiplies the data by a codeword resulting in a longer data sequence. It is used where there is insufficient signal level to support communications using non-spread communications. Broadly, it trades off extra range against a reduced data rate.
- *Cyclic prefix insertion.* This adds a repetition of the end of the frame to the start of the frame. This allows the received frame to be readily converted into the frequency domain uncontaminated by multipath from previous transmissions.
- *Synchronisation (sync) insertion.* This adds known patterns of bits that can be used by the receiver to synchronise its internal clocks to the transmitter.
- *Root raised cosine (RRC) pulse shaping.* This turns the square wave binary signal into a more sinusoidal pulse to reduce out-of-band emissions when transmitted.

Note that interleaving is not used as most bursts of data are too short for it to bring benefit. Generally, interleaving is used to distribute errors more evenly across received data. In a radio system errors can tend to occur in clusters when particularly bad propagation conditions are experienced. These clusters can overwhelm the error correction system so interleaving attempts to distribute them across multiple error correction blocks so that all blocks experience a similar error rate, ideally within the capabilities of the error correction system to correct. However, if a terminal only

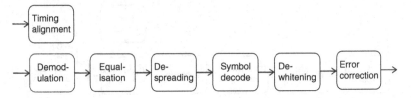

Figure 7.2 Receive process

transmits a small number of blocks of data the room for such interleaving to spread clusters of errors is limited.

Next the signal is converted to radio frequency (RF). Frequency hopping is employed so the frequency that the signal is converted to will vary from frame to frame.

The receiver broadly follows the reverse process as shown in Figure 7.2.

The key differences are that:

- Symbol decoding is used to turn the shaped signal back into binary data.
- The inserted synchronisation sequence is used to achieve timing references.
- The cyclic prefix is used in the equaliser that compensates for frequency selective fading in the channel.

One of the key features of the PHY layer is that it accommodates a very wide range of path loss values corresponding to terminals close to the base station to some distance away – perhaps 10 km. An overview of how it does this is provided in Table 7.1.

We will return to discuss this table in more detail at a later stage, the key point to note here is that the PHY layer uses a combination of the number of modulation levels (QAM to QPSK to BPSK to DBPSK), the error correction rate (1 to 3/4 to 1/2 rate) and the spreading factor (1 to 1023) to accommodate received signal levels varying from −82 dBm to −128 dBm. Broadly, the process followed as the signal level falls from the highest to the lowest received levels is:

Table 7.1 Overview of variation in PHY parameters [Source: Weightless standard]

Modulation scheme	Coding rate	Spreading factor	Downlink PHY data rate (Mbps)	Required SNR before FEC & spreading (dB)	FEC gain (dB)	Spreading gain (dB)	Required SNR for 10^{-4} BER (dB)	Noise figure incl. digital losses (dB)	Required signal level at Rx input (dBm)
16-QAM	1	1	16.0	18.5	0.0	0	+18.5	6.0	−82.5
16-QAM	3/4	1	12.0	18.5	4.0	0	+14.5	6.0	−86.5
16-QAM	1/2	1	8.0	18.5	7.5	0	+11.0	6.0	−90.0
QPSK	3/4	1	6.0	11.5	4.0	0	+7.5	6.0	−93.5
QPSK	1/2	1	4.0	11.5	7.5	0	+4.0	6.0	−97.0
BPSK	1/2	1	2.0	8.5	7.5	0	+1.0	6.0	−100.0
BPSK	1/2	4	0.5	8.5	7.5	6.0	−5.0	6.0	−106.0
BPSK	1/2	16	0.125	8.5	7.5	12.0	−11.0	6.0	−112.0
BPSK	1/2	63	0.040	8.5	7.5	18.0	−17.0	6.0	−118.0
BPSK	1/2	255	0.010	8.5	7.5	24.0	−23.0	6.0	−124.0
DBPSK	1/2	1023	0.0025	10.5	7.5	30.0	−27.0	6.0	−128.0

- At the strongest signal level use the highest modulation level and no error correction.
- Then add increasing error correction.
- Then step down a modulation level but reduce the error correction.
- Repeat the above two steps until the lowest modulation level is reached (BPSK).
- Then progressively add spreading.
- Finally, move to differential modulation.

Note that the modulation scheme used is a form of single-carrier modulation rather than the code division multiple access (CDMA) used in 3G or orthogonal frequency division multiple access (OFDM) used in 4G. The reason for not using CDMA is that it requires accurate timing and power control of the messages from terminals but with many messages being short bursts there is insufficient time for power control loops to settle. This could result in a significant drop in capacity. OFDM primarily simplifies the equalisation problem for very high data rate transmissions. Since data rates in Weightless are low compared to 4G, OFDM is not needed and simpler approaches can be adopted. Further, OFDM requires a high peak-to-average power ratio which results in relatively high battery drain. With long battery life a key concern within Weightless this is a significant disadvantage. Other factors in the design choice also included a desire to keep terminal royalty costs as low as possible by avoiding technologies where there were known to be significant intellectual property right (IPR) portfolios. Many of the core techniques used in Weightless have been known for more than 20 years and hence are no longer covered by valid patents.

Having understood the interaction between modulation, error correction and spreading we now look at each part of the PHY process in more detail.

7.3 Forward error correction (FEC) encoding

FEC is implemented using convolutional codes. These work by passing the input through a shift register and then using a Viterbi decoder to determine the most likely sequence transmitted. The basic encoder is a 1/2 rate (i.e. for every data bit an encoded bit is added). The 3/4 rate encoding is achieved by puncturing the 1/2 rate coding – effectively removing every 4th bit from the sequence. Padding of 7 bits of 0 is added to the transmitted sequence to allow the receive decoder to 'settle'. Full details of the polynomial used to implement the shift register are provided in the *standard*.

The optimal level of error correction to use is a balance between the redundancy caused by adding the coding information or the loss of capacity in using fewer modulation levels or longer spreading codes, and the loss of capacity caused when bursts received in error need retransmission. Simulations of Weightless systems suggest that the optimal point is reached when bursts have an error rate of between 5% and 10%. That is, the CRC check on a burst fails between 5% and 10% of the time resulting in the burst requiring re-transmission. The network selects the coding scheme (and modulation levels and spreading factor) to try to result in the burst error rate falling within this window. However, it may be that the channel changes from the time that the network sets the coding scheme and the transmission takes place, in which case the network will further adapt the coding scheme for subsequent transmissions.

7.4 Whitening

Whitening is achieved by XORing the data sequence with a sequence that approximates to white noise – i.e. it has a fairly random distribution of 1's and 0's. The whitening sequence is generated by passing a seed based on the frame number into a shift register. In the case of the DL_FCH the seed is based on the channel number on which the frame is being transmitted.

Basing the sequence on the frame number or channel number is beneficial for the following reasons:

• It ensures that retransmitted bursts following a CRC error undergo different whitening compared with the original transmission (because they will be in a subsequent frame). This means that should a data sequence have been generated originally that caused particular problems for the receiver, it will not be generated again on re-transmission.

- It ensures that a receiver tuned to one channel is very unlikely to decode a valid FCH field from a transmitter operating on a different channel which might 'accidentally' be received due to image frequency issues within the receiver.

Details of the shift register polynomial and seed manipulations are provided in the *standard*.

7.5 Modulation

The modulation is standard use of 16-QAM (4 bits/symbol), QPSK (2 bits/symbol) or BPSK (1 bit/symbol). A $\pi/2$ rotation per BPSK chip, or $\pi/4$ rotation per QPSK chip, is applied in order to reduce the peak-to-average power ratio (PAPR) of the modulation.

7.6 Spreading

Each bit of the data sequence is multiplied by the spreading code. If the code is of length n this increases the length of the sequence by a factor of n. As mentioned above n could be relatively small such as 15, or very large, up to 1023. The effective impact is that if the data bit is '1' then the spreading code is transmitted whereas if it is '0' then the inverse of the spreading code is transmitted.

7.7 Cyclic prefix insertion

The cyclic prefix is a repetition of data from the end of the frame to the start of the frame. In essence, it allows frequency domain equalisation to be performed at the receiver by taking a fast Fourier transform (FFT) of the received block. How this is done is discussed in more detail below.

The transmission is partitioned into blocks of chips, with each block having a length of 1024 chips and a cyclic prefix of 144 chips and postfix of 16 chips.

The cyclic prefix length has been chosen to exceed the maximum channel delay spread. The cyclic postfix length is related to the impulse response length of the RRC pulse shaping filter to avoid inter-block interference due to the pulse shaping.

The use of the cyclic prefix and postfix simplifies the equaliser implementation at the receiver by allowing the use of a frequency domain equaliser, analogous with that used in an OFDM receiver. The overall modulation scheme is commonly called 'single-carrier, frequency domain equalisation' (SC-FDE) [1]. It has many of the benefits of a multi-carrier OFDM scheme, particularly in terms of reduced complexity equalisation for long delay spread channels, but it avoids the disadvantage of high PAPR in an OFDM system.

Transmissions are padded to form an integral number of blocks. The data used for padding are zero bits at the input to the whitener. The length fields within the MAC burst headers specify the number of data bits excluding the padding.

At the receiver, the data is segmented into blocks starting with a prefix and ending with a postfix. The prefix is then discarded. This is because it can contain multipath signals related to information transmitted in the previous block. (The multipath from the prefix which falls beyond the prefix relates to information transmitted towards the end of the frame, so all multipath energy in the block now relates to data from that block.) The block is then transformed into the frequency domain using an FFT.

The receiver also makes use of the synchronisation sequence to 'sound' the channel. As well as gaining timing information, the synchronisation sequence allows the receiver to determine the various multipath elements, characterising each path in terms of delay, attenuation and phase shift (in the time domain). One option is then for the receiver to convert this response to the frequency domain by undertaking an FFT. The equaliser coefficients are then formed as the inverse of this frequency response. However, where the frequency response falls close to zero it can lead to very high equaliser coefficients which would magnify noise so some co-efficient weighting is performed to reduce such elements.

Equalisation can then occur in the frequency domain by multiplying the signal by the inverse FFT of the channel response – effectively a frequency domain equaliser. Finally, the signal is transposed back to the time domain using an inverse FFT (IFFT).

7.8 Synchronisation (sync) insertion

A synchronisation sequence is transmitted prior to each burst. This is used for:

- detection of the start of a burst
- fine frequency error estimation
- channel estimation
- timing detection for the start of the payload.

The synchronisation sequence is composed of repeated blocks of length 128 chips, followed by three termination blocks that are inverted. The synchronisation blocks are derived numerically for flat power spectrum and low PAPR. The repeated nature of the synchronisation sequence supports frequency domain correlation as an alternative to time domain correlation. The termination blocks allow detection of the end of the synchronisation sequence.

If the SNR is low then more synchronisation blocks are required by the receiver. For spreading factors 1 to 32 then 26 blocks are sent, for factor 32 then 36 blocks are sent and for factors greater than this 72 blocks are sent.

Base stations may use different synchronisation sequences in order to reduce the likelihood of a terminal obtaining a false synchronisation based on a transmission from a more distant base station. The assignment of sequences to the base stations is performed by the network as discussed in Section 5.4.

7.9 Root raised cosine (RRC) pulse shaping

The binary data is passed through an RRC filter with $\beta = 0.4$ roll-off. This results in a smooth pulse shape that minimises unwanted emissions.

7.10 Frequency hopping

Frequency hopping is applied at the frame rate as discussed in Section 5.3. The frequency hopping sequence is determined by the network and sent to the base stations. As well as providing mitigation against interference from other users of the spectrum it prevents the terminal being in a deep fade for a long duration.

7.11 Downlink, uplink and TDD

Weightless transmissions are time division duplex. The downlink is transmitted followed by the uplink. There are multiple types of downlink and uplink as follows:

1. High data rate downlink. This provides data rates in the region of 16 Mbits/s to 500 kbits/s.
2. Low data rate downlink. This provides rates from 500 kbits/s to 2.5 kbits/s.
3. Narrowband uplink. This is the preferred solution.
4. Wideband uplink. This can be used in the US to best meet regulatory requirements.

The manner in which each of these differs is described below.

7.11.1 High data rate downlink

This makes use of the approach described above. It only uses spreading factors up to 15.

7.11.2 Low data rate downlink

This uses spreading factors of 32 and above. Such long spreading factors allow for equalisation using a rake receiver. This uses correlation of the transmitted codeword to measure the multipath signals received which can then be used to set the equaliser coefficients. This means there is no need for the cyclic prefix. At these high spreading factors QAM is not used as there is insufficient SNR.

Data bits are mapped onto modulated signals through a combination of:

- BPSK and QPSK modulation, applied coherently or differentially (the differential forms of these modulations are referred to as DBPSK and DQPSK)
- Single-code and multi-code modulation

Differential modulation is where the information is encoded in the difference between one modulated symbol and the next rather than the absolute value. It is simpler to decode as the receiver does not need to obtain accurate frequency and phase information. However, it tends to result in a 3 dB penalty in that if one symbol is decoded in error then the difference from the previous symbol to this symbol and from this symbol to the next symbol will both be incorrectly estimated resulting in two errors rather than the single error that would have occurred with coherent modulation. Hence, it is only used in the lowest signal level case where reception using coherent (non-differential) modulation would typically not be possible.

The basic modulation scheme is single-code. A fixed spreading code is used for all symbols in the burst, and so data bits are encoded using phase modulation only. Multi-code modulation allows additional information to be encoded by means of code selection. The transmitter selects 1 of C possible spreading codes to convey $\log_2 C$ bits of information, in addition to any information that is encoded based on phase modulation. The maximum supported value of C is 4, so this implies that the receiver must correlate against a maximum of four possible codes.

The use of multi-codes provides two main benefits:

- Improved performance, which may be viewed as a form of coding gain. This benefit is reduced in the presence of FEC but can still be advantageous.
- A given data rate can be achieved with a larger spreading factor. Increasing the spreading factor reduces the data rate because more chips are sent per data bit. But using multi-codes increases the number of encoded bits per symbol due to the extra dimension introduced

hence allowing a given user data rate at a greater spreading factor than would be the case without multi-coding. Extending the spreading factor improves the ability of the rake receiver to remove the multipath when operating with higher delay spread channels.

A special case of multi-code modulation is when no phase modulation is utilised such that all the encoded data is by means of the code selection. This allows the receiver to perform non-coherent detection, which minimises the impact of low-frequency phase noise at the expense of degraded performance. It is therefore most applicable at the highest spreading factors. This mode may be viewed as a form of Orthogonal Shift Keying (OSK) since the spreading codes are approximately orthogonal.

7.11.3 Narrowband uplink

Weightless terminals are likely to transmit with much less power than Weightless base stations. This is both because the base stations will have mains power and also due to regulatory constraints in white space that tend to restrict mobile devices to lower power levels. However, it is important to have a balanced link budget. Weightless achieves this using a narrower band uplink than downlink. The narrower band has a lower noise floor compensating for the lower received signal strength. Multiple narrow band uplinks can then be accommodated in one white space channel – hence the uplink makes use of FDMA within the channel whereas the downlink is a single carrier within this channel.

For the uplink it may be necessary to implement power control to reduce the power of some or all terminals in a cell. Power control is needed for two reasons:

- To reduce the dynamic range requirements on the base station receiver by reducing the transmit power of terminals that are close to the base station.
- To limit the transmit power of all terminals in a cell because access to certain white space channels may only be allowed for reduced terminal transmit power.

Fast or finely quantised power control is not required, unlike in a CDMA system. This is because separation of users is achieved though frequency division rather than code division, and so the system is not sensitive to sub-optimum cross-correlation properties between spreading codes. Therefore, power control is achieved though relatively high latency MAC control messages sent to individual terminals or broadcast to all terminals.

The various possible encoding schemes and resultant performance are shown in Table 7.2.

7.11.4 Wideband uplink

The FCC regulations for white space access measure transmitted power in a narrow 100 kHz bandwidth rather than across the entire 6 MHz white space channel. This penalises narrowband transmissions which need to use a much reduced power level compared to broadband transmission. To compensate for this, Weightless has a wideband uplink mode where instead of a single FDMA channel a comb of channels is produced. This is shown in Figure 7.3.

Separate terminals produce combs offset slightly from each other so they do not overlap. Using this approach, a low power level is produced in any 100 kHz bandwidth but the full channel allocation can be reached. This overcomes the rather poorly designed US regulations but at the cost of increased complexity and a slight reduction in capacity – only 16 uplink channels are available compared to 24 for the narrowband uplink. Hence, this mode is not preferred and likely only to be used in the US and any other countries that implement similar regulations.

The wideband uplink makes use of all the features of the narrowband uplink described above.

7.12 RF performance

The performance requirements of Weightless are detailed and provided in the *standard*. It is important that terminals achieve a minimum RF performance for the following reasons:

Table 7.2 *Uplink coding schemes and performance [Source: Weightless standard]*

Modulation scheme	Coding rate	Spreading factor	Uplink PHY data rate (Mbps)	Required SNR before coding & spreading (dB)	FEC gain (dB)	Processing gain incl. sub-channel and dual-Rx (dB)	Required SNR for 10^{-4} BER (dB)	Noise figure incl. digital losses (dB)	Required signal level at Rx input (dBm)
16-QAM	1	1	**0.500**	18.5	0.0	18.0	**−0.5**	6.0	**−100.5**
16-QAM	3/4	1	**0.375**	18.5	4.0	18.0	**−3.5**	6.0	**−104.5**
16-QAM	1/2	1	**0.250**	18.5	7.5	18.0	**−7.0**	6.0	**−108.0**
QPSK	3/4	1	**0.187**	11.5	4.0	18.0	**−10.5**	6.0	**−111.5**
QPSK	1/2	1	**0.125**	11.5	7.5	18.0	**−14.0**	6.0	**−115.0**
BPSK	1/2	1	**0.063**	8.5	7.5	18.0	**−17.0**	6.0	**−118.0**
BPSK	1/2	4	**0.016**	8.5	7.5	24.0	**−23.0**	6.0	**−124.0**
BPSK	1/2	16	**0.004**	8.5	7.5	30.0	**−29.0**	6.0	**−130.0**
DBPSK	1/2	64	**0.001**	10.5	7.5	36.0	**−33.0**	6.0	**−134.0**

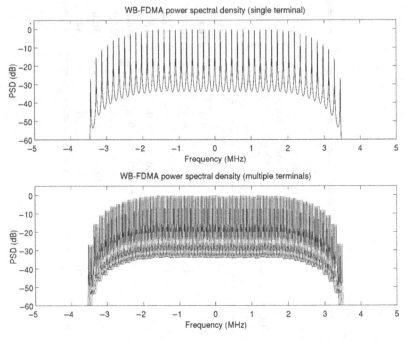

Figure 7.3 Wideband uplink frequency response [Source: Weightless *standard*]

- Some regulatory approaches, such as that in the UK, use the adjacent channel performance of the devices in determining white space availability. If terminal adjacent channel performance is poor this could reduce white space availability across the whole network. In addition, since the terminal performance needs to be provided to the regulator, there needs to be a mechanism to ensure all terminals can achieve this.
- If the terminal sensitivity is poor (it cannot receive weak signals) it will tend to operate at a higher spreading factor. As explained in the next chapter, this can have a major impact on the overall network capacity.
- If there is not some common understanding of the performance of the terminals and the base stations then some terminals may not work effectively with some base stations.

Many parameters, such as frequency accuracy, are not appropriate for this book. However, it is worth explaining the out-of-band performance as this relates to white space activities.

The primary aim of the transmit mask is to ensure that out-of-band emissions do not interfere with licensed users of the band. For many white space applications the licensed users who will require protection are terrestrial digital TV receivers. Details of the protection ratios needed for typical DTV receivers have been provided by some regulators. For example, Ofcom specify that a receiver requires 20 dB C/I on channel and -30 dB on the adjacent channel. This suggests that if a Weightless device operating on the adjacent channel to a TV receiver sets its emissions into the TV band at 50 dB (20–30) down from its in-band emissions then the impact of its emissions within its allowed band and the TV receiver band will be identical. Hence, if the adjacent channel emissions for Weightless are greater than -50 dBc (dB relative to the carrier) these will dominate interference to the TV. If they are less than -50 dBc then the interference from the Weightless carrier will dominate. US specifications require -55 dBc which appears to be excessively cautious but at the time of writing there was some indication these might be relaxed through making the limit an absolute power level rather than relative to the carrier.

Hence, Weightless devices should achieve around -50 dBc on the adjacent channel. More details of how the full spectrum mask is derived are provided in the *standard*.

7.13 Antenna issues

While many new cellular and Wi-Fi technologies have been opting for multiple antenna schemes such as multiple-input multiple-output (MIMO) antennas, Weightless has deliberately avoided these. This is because:

- Weightless systems will generally not be as constrained regarding capacity as cellular systems so the extra throughput is less important.
- MIMO works best with high SNRs but many Weightless terminals will be operating with low SNR and high spreading factors. As

discussed in Chapter 9 it is these low SNR terminals that dominate capacity and hence implementing MIMO for the high SNR terminals would bring very little benefit.

- Terminals need to be very low-cost. Multi-antenna systems add cost through the need for multiple antennas and more powerful processing needed. This can also increase power consumption, reducing battery life.
- MIMO systems are often covered by IPR potentially adding royalty cost to the terminals.

Weightless assumes that there is only a single antenna at the terminal. However, multiple antennas can be adopted at the base station and could be used in a number of ways.

- To steer a null towards a source of interference, as explained in Section 5.5. At least two antennas are needed to achieve this.
- To implement maximum ratio combining on the uplink, combining the received signal at both antennas to overcome fades that might be occurring at one of them and increasing the overall received power.
- In principle, simultaneous transmission from both antennas would be possible to improve downlink data rates but there is no intention to implement this at the time of writing. This is because it would result in increased terminal complexity.

As a result, it is expected that most base stations will use two antennas and would select between null steering and diversity reception depending on the strength of any interference in the cell. Some may use more antennas where space allows for further improved performance. Antennas can be simple dipoles of about 0.5 m height or alternatively stacked dipoles can be used resulting in a large antenna (e.g. 1.5 m for a three-dipole array) but providing additional gain as a result of the narrower vertical beam pattern. Selection of the number and type of antenna is for the operator.

7.14 The attach process as an example

7.14.1 Introduction

To conclude the chapters on MAC and PHY a worked example of the actions a terminal would take on first being switched on are provided.

This is intended to illustrate a number of points and demonstrate how terminal design can influence performance.

In overview the process is:

1. Discover a channel that the network is using.
2. Decode the FCH to understand network parameters.
3. Send a contended access message to attach.
4. Undertake authentication.
5. Agree a schedule for the next transmitted message if appropriate.

Each of these stages is described below.

7.14.2 Channel discovery

When a terminal is first powered up it has no understanding of what cells cover its location or what frequencies they are using. The frequency hopped nature of transmissions makes acquisition particularly complex.

The simplest approach for the terminal is to dwell on each possible white space channel for long enough to determine whether it is used as part of the hop sequence. With a maximum hopping sequence of 8 and a typical frame length of 2 s a terminal would need to dwell on a channel for 16 s to determine whether it can detect a valid synchronisation sequence. It would then move through all white space channels of which there are typically around 32 available depending on the local regulations. This would take around $8\frac{1}{2}$ minutes. Note that it should continue to scan available channels even when it does discover a synchronisation sequence as it is possible that it has heard one from a neighbouring cell rather than the cell that provides optimal coverage for its location. Once it has concluded on the channel with the strongest signal it can return to this channel to re-await the synchronisation sequence and move to the next step.

The *standard* does not provide details on acquisition strategies as that is left to manufacturers. However, faster strategies could be envisaged. For example, the device could quickly scan all available channels in a short time period (of the order 1–2 s). On each channel it could look for transmission activity that had a spectral characteristic like that of

Weightless. It could perform this scan a few times to overcome the risk that a base station was on its uplink cycle when the measurement was performed. It could then select the channel with the strongest signal and attempt decoding of the synchronisation burst. Such an approach could be completed in only 10–15 s but might not work depending on the strength of signal from the base station.

A terminal might also adopt an approach of synchronising with the first channel it finds and attaching to that base station in order to minimise the acquisition time where this is a concern to users. Then, in a time period when it is idle it could complete a full scan and attach to a different base station if it provided a stronger signal level. Situations where this might be relevant include the installation of a smart meter where the installer might wish to verify quite quickly that the meter was working correctly and then subsequently the meter could optimise its connection. Application areas are discussed further in Chapter 10.

7.14.3 Decode the FCH

The terminal then synchronises using firstly the frame synchronisation burst to obtain coarse timing and then the synchronisation sequence within the FCH to obtain fine timing. It is now able to determine the following:

- Whether the base station belongs to a network that it has access rights for.
- The frame number and hence the position in the super-frame.
- Any transmit power restrictions that apply.
- The frequency hopping sequence for the cell.
- The parameters relevant to decode the RS_MAP.

In the unlikely case that the frame is a broadcast frame then the terminal will listen to the broadcast information as needed to understand whether parameters such as the hopping sequence are about to change.

7.14.4 Send a contended access message to attach

The terminal next needs to read the RS_MAP to determine what contended access resource is available within the frame. It then follows the CA process, generating a random seed and backing off as needed. (Back off might seem unnecessary for an initial attach but it is possible that many devices in a cell are powered off by some event and powered back up simultaneously and all attempt to attach.) It sends its CA message either with its UUTID if there is room or a random number otherwise and follows the process described in Section 6.7 to obtain an iTID and register on the network.

7.14.5 Undertake authentication

Once the network has the UUTID of the terminal it can move to authenticate it. It does so by sending the terminal a message encoded using the one-time pad that both it and the terminal have stored securely. The terminal decodes this message and responds by sending a known message back that is also encoded. If this process completes satisfactorily then both parties are authenticated.

7.14.6 Agree a schedule for the next transmitted message

The terminal then has the opportunity to send further information to the base station. This may be some initial message it wishes to send. Finally, the base station will inform the terminal of its next assigned frame for communications, if relevant. For example, for a smart meter it may provide it with pre-assigned resource. The terminal can then return to idle mode until it needs to either listen to a broadcast frame or use its pre-assigned slot. During this idle time it may decide to perform other tasks such as scanning to see if there are better base stations to attach to.

Reference

[1] See www.utdallas.edu/~aldhahir/fde_final.pdf and the references contained within

8   Further functionality

8.1 Encryption and authentication

8.1.1 Security in general

Security is important in many machine applications. Smart grids must return accurate meter readings that cannot be modified. Network communications to instruct meters to change demand must not be modified or sent by a terrorist organisation. Financial-related transactions must be secure and it must be certain that they were received. Almost all machine communications have some level of security requirements.

To some degree these could be layered over the top of the Weightless network at the application layer. So a smart meter could encrypt its data stream before passing it to the Weightless radio embedded within it. The data would pass through the Weightless network in secure form and be decoded by the client in their central IT system. Indeed, this can be done even when Weightless has its own security to provide extra protection.

There are some functions that cannot be achieved at the application layer. These include authentication of the network by the Weightless device. It would be possible for a terminal to be 'captured' by a rogue base station and stay attached to it indefinitely. Authentication mechanisms are needed to prevent this occurring. Because of this, and because it prevented all applications having to build in their own security, it was decided to provide the Weightless standard with cellular-grade security mechanisms. However, as will be explained, some of the differences in message type and threat type mean that the same security mechanisms as are used in cellular cannot just be copied.

There are three security-related functions relevant to machine networks:

- Authentication – the ability of one end of the link to determine that the other end is the device or base station it claims to be.
- Encryption – making it near-impossible for anyone else to read messages.
- Repudiation – being able to prove that a message was delivered as claimed.

Security against other threats follows from this. For example, a 'replay attack' where a message is recorded by a third party and replayed cannot succeed if there is authentication.

8.1.2 Security for machine communications

Machine communications have some important differences to personal communications. Messages tend to be much shorter. They can also be near-repetitive, for example sequential meter readings may only differ in the least significant digits. Repetitive messages provide strong opportunities for eavesdroppers to crack some codes. In general, authentication is more important than encryption for machines – the opportunity for stealing energy are greater by impersonating meters than they are through listening to the readings from a neighbour's meter (which may be in an accessible cabinet on the outside of the house in any case).

8.1.3 Weightless security

Weightless security was one of the least developed areas in the v0.6 of the *standard*. The intention was that the security sub-group would bring together a wide range of expertise allowing this area to be better developed than would be the case were a single company to determine the approach. Hence, at the time of writing, there was very little information on security.

In the draft *standard* a tentative decision was taken to combine authentication with encryption. When a terminal attaches and provides its full identity it is sent an encrypted message. If the terminal can correctly decode this message it knows that the network must be valid since an invalid network would not know how to formulate a correctly encoded message. It then responds to the network with a known response encoded using its key system. The network then knows that the terminal is the one it claims to be as no other terminal would be able to encode the message correctly. Hence, both the network is authenticated to the terminal and the terminal to the network.

For subsequent communication the encryption system is used, providing secrecy. Acknowledgements provided during the communication process form the proof of delivery and can be stored indefinitely either in the Weightless network or by the client preventing repudiation.

The scheme designed by the sub-group should protect identity, provide secure mutual authentication, secure session key establishment, work well with a minimum number of packet exchanges and be protected against future developments in computation and cryptography.

At the time of writing, the precise details of the Weightless security algorithm were still under development.

8.2 Moving base stations

There has long been an idea that meters could be read by a van driving up and down the street. Meters would have a short-range wireless system which would wake up when the van came within range and transmit their reading.

The widespread coverage expected from Weightless systems generally renders this unnecessary. However, there may be situations in very remote areas where it is not economic to provide coverage. In these cases, Weightless supports the concept of moving base stations that can be used to periodically provide coverage. Note, this would be of little value for a personal communications system where coverage is needed when the user wishes to interact, but is perfectly useful for machines that can buffer any data to be sent until they have communications facilities available to them.

The key difficulty with mobile base stations is acquisition of the base station by the terminals. The terminals need to wake up, scan available channels, find the base station, then transmit a message all before the base station passes out of range. The time available to do this will be dependent on many factors such as the speed of the mobile base station, how closely it passes to the terminal and its transmit power. The faster that terminals can acquire the base station the more likely it is that this mode of operation will work.

However, rapid acquisition requires frequent scanning for the presence of a channel which requires high power consumption by the terminal. In some cases, such as where the terminal has an external power supply, this may not be an issue, in other cases it might severely limit the frequency of scanning that the terminal can support. Moving base stations indicate that they are mobile in their header message and terminals that make use of them will need to decide how to behave, making their own trade-off between scanning frequency and power consumption. In some cases, it might be necessary for the terminals to be pre-programmed to look for moving base stations such that they search appropriately from the moment they are initialised. How this is done is outside the scope of this specification. Once a terminal has acquired a moving base station it should assume that it will need to operate via moving base stations until such time as it acquired a standard base station.

In order to assist acquisition, moving base stations might opt not to use frequency hopping and to use the same channel or sets of channels each 'visit' (although the ability to do this depends on the white space availability remaining unchanged). Terminals that have acquired the moving base station may choose to optimise their acquisition/energy usage mix by monitoring the channel(s) last used by the moving base station more frequently than other channels.

Moving base stations are assumed not to have backhaul connectivity while travelling and so need to operate autonomously. This raises a number of issues as discussed below.

Moving base stations need to obtain white space availability for the area they will cover with their transmissions while en route and ensure that this availability is valid for the duration of their trip. The base station may need to switch channels en route in some cases. The moving base station will need some understanding of both its current location and the coverage it is generating from that location.

The moving base station will also need much of the functionality normally resident within the network including frame assembly. Hence, it might be considered more of a 'moving network' than a 'moving base station'.

There are a number of options as to how encryption might be achieved:

1. No encryption is adopted for mobile base stations (not recommended).
2. Messages for the terminal are encrypted by the network before being pre-stored in the mobile base station and messages from the terminal are stored in encrypted form in the mobile base station before being passed back to the network on return. This preserves the full level of security but prevents 'unexpected' exchanges of information with the terminal.
3. The mobile base station can be given the encryption key set for the terminals it will encounter (or the complete set for all terminals). This provides flexibility but risks compromising network security unless the mobile base station can be fully trusted by all parties.

The choice between these different options is for the network operator.

The moving base station will need to update the network appropriately on return with details such as billing information, updates to the location registers and to the customer portal as appropriate.

8.3 Location measurement

Measurement of location is not a core part of the Weightless system but is provided as an additional 'feature'. In many cases terminals will either be pre-populated with their location or will have available accurate systems such as GPS. If not, a Weightless network may be able to locate the terminal itself depending on factors such as how many base stations cover the area where the terminal is located and the accuracy of timing synchronisation across the network. Weightless supports three mechanisms, the selection of which is up to the network operator. At the time of writing the exact details of each scheme were still being studied. The schemes are:

1. Passive-terminal.
2. Passive-network.
3. Active.

In passive-terminal a terminal monitors the transmissions from as many base stations as possible and determines the relative time of arrival of the synchronisation burst from each. It also decodes the base station identity of each. It then sends to its 'home' base station the identities and timing relative to the home base station of each. The base station then passes this information to the location server in the network which uses its knowledge of the location of each base station to triangulate the terminal location. This is then sent back to the client (but not to the terminal which typically does not need to know its own location). Passive location requires accurate synchronisation of each of the base stations implying GPS timing or similar. The terminal needs to understand the actions it must perform when requested to make such a measurement and the network needs a location engine that can perform the triangulation. Support of this feature in a terminal is optional.

In passive-network base stations surrounding the cell where the terminal is located listen for transmissions from the terminal. They pass details of the timing of the received signal back to the location server. This is not a preferred solution as it typically requires additional receivers in the base station able to scan other channels than the one that the base station is currently using for transmission and reception in its cell. This would add extra cost to the base station. It does not require any special action on behalf of the terminal.

In active location measurement the terminal detaches from the home base station and attempts to attach to as many other base stations as it can. Once attached the terminal requests a timing measurement be made. The base station then notes the difference in timing between the start of an uplink frame and the actual time that the terminal transmission is received and forwards this information and the terminal identity to the location server. Once the terminal has completed this process it returns to the home base station and reattaches. The network then signals to the location server that the process is complete and the location server deduces the terminal location. This process does not require that different base stations are synchronised but is more time consuming and also uses some network resources. It requires that the terminal understand the actions required of it and may not be supported in all terminals. It can be the most accurate of all methods since excellent synchronisation is obtained with each base station and round-trip delay can be accurately measured.

8.4 Roaming

It is unclear how the network deployment and ownership structure for Weightless networks might evolve. There could be:

- One national network per country providing ubiquitous coverage.
- Multiple national networks, each providing good coverage, as is the case with cellular networks.
- Multiple regional networks.
- Privately owned networks covering small areas such as an oil refinery.
- Privately owned base stations with coverage and capacity 'sold' into a core network.
- Any combination of these, changing and evolving over time.

Terminals may move between these networks and hence wish to register onto different networks.

When terminals search for networks they decode the base station identity and based on the upper few digits determine which network it belongs to. If the base station does not belong to their home network then they continue to scan other available base stations and if they discover a home network base station they attach to that base station even if it does not have the strongest signal level.

If terminals are unable to find a home network they may consult an internal list of allowed visited networks and select the network to attach to in accordance to this. Alternatively, they may simply attempt to attach to the local network.

When a network receives an attach message it can look up the UUTID to see if that terminal 'belongs' to its network. If not, it can send the UUTID to a global subscriber database to receive information as to the 'home' network for that UUTID and any rights it has to roam onto other networks. Based on this, it can decide whether to allow the terminal access to its network. If it does, it proceeds as per a normal attach process. If not it sends a detach message to the terminal with cause 'network not allowed'. The terminal should not attempt to attach to that network again for a pre-defined period (e.g. at least 1 day).

After a successful attach the roaming network sends a message to the location register of the home network. The home network will then update the location register with details of the visited network as appropriate.

When roaming the authentication/encryption function remains with the home network – that is the ciphering key is not passed to the visited network, instead information for the terminal is encrypted prior to being forwarded to the visited network.

Where there are overlapping networks there is a risk of interference if both use the same frequencies at the same time. This is no different to any other unlicensed user transmitting and the use of techniques such as frequency hopping will mitigate the impact of this interference. However, improvements can be made on this. The first step is to synchronise the networks such that the frames start at the same time. This will tend to keep downlink and uplink transmissions approximately aligned, minimising the impact of base station to base station interference. The next step is to engage in joint frequency planning such that hopping sequences are devised that minimise interference both within and across networks. This adds complexity to the frequency assignment algorithm but it can still operate in predominantly the same manner as before. Sharing frequency planning implies sharing information across the core network. Since the core networks for the two different networks are both virtual machines the obvious next step is to combine them. Taken to its logical conclusion, this suggests a single core network for a country or region with separate operators building the base station infrastructure.

9 Network design and capacity

9.1 Understanding the capacity of a cell

The headline data rate of Weightless is 16 Mbits/s. However, the capacity of a cell is typically far from this. Firstly, Weightless is a TDD system. If the time is divided half between the uplink and the downlink the net maximum capacity would be 8 Mbits/s symmetrically. But the main reduction in capacity comes from the fact that terminals can rarely access the highest data rates. As shown in Table 7.1 terminals step down through modulation schemes and then to increased spreading factors as they move further away from the base station. This progressively reduces the data rate available on the downlink from 16 Mbits/s right down to 2.5 kbits/s. So, simplistically, in the highly unlikely case that all the terminals in the cell were at maximum range and the cell was split 50:50 downlink to uplink then the capacity would be 1.25 kbits/s symmetrical. Happily, much higher rates than this can be achieved in practice!

Typically terminals will be distributed around the cell. The actual distribution will vary dramatically depending on the circumstances – for example in a cell in the centre of a town, most smart meters might be distributed close to the base station with few at cell edge. The capacity of the cell is then some average of the data rates achieved by each terminal.

It is worth considering how this averaging is performed. One approach would be to give all terminals equal transmission time. This would mean those close to the base station could transmit more during this time than those further away. This is the averaging approach often used in cellular systems and it tends to result in a high average because those close in get extremely high data rates. However, most machines have fixed amounts of data to send, such as a meter reading. Having higher data rates available to them will result in this being sent more quickly. Hence, the averaging needs to be based on equal resource, that is giving each terminal the resource it needs to send a message of a certain size. In such an approach, the capacity tends to be dominated by the terminals furthest out because of the long time they need for transmissions. The results are much lower than for 'equal time' averaging and appear to compare badly with cellular systems. However, they are the only results that are relevant for machine communications and so are the ones provided here.

Calculating the cell capacity is then a case of:

- determining where the terminals are in the cell;
- determining the path loss from the base station to each terminal;
- determining the resulting modulation scheme and coding rate used for that terminal and hence the data rate;
- working out the time taken to transmit a bit of information;
- taking the average of this time across the cell;
- dividing the average time per bit into 1 s to get the average throughput.

For the initial modelling discussed in the section (more complex modelling is presented in later sections), we have assumed:

- a uniform spread of terminals across the cell
- use of the Hata model to predict path loss
- the data rates specified in the Weightless standard

The resulting data rates versus cell radius are shown in Figure 9.1 for a 50:50 TDD split (all subsequent results will be for this TDD split unless stated otherwise). Other assumptions are 8 MHz channel bandwidth and 36 dBm EIRP transmit power.

It is clear (and unsurprising) that the cell capacity increases as the cell gets smaller. This is because more terminals are closer to the base station and so have higher data rates, requiring less time to send their messages. In fact, the impact on overall network capacity of small cells is even more extreme as shown in Figure 9.2.

As more cells are put in two factors occur. The first is that each cell adds extra capacity – replacing one large cell with four small cells immediately brings a four-fold increase because more cells allow reuse of channels. This is shown by the dashed 'extra cells' line in the figure.

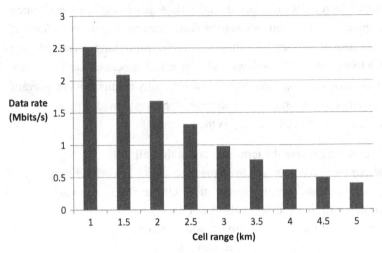

Figure 9.1 Cell capacity for a single Weightless cell [Source: Neul]

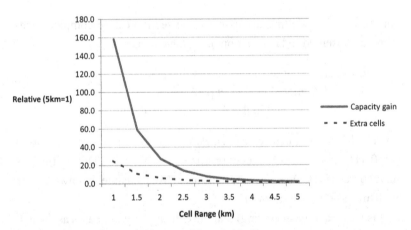

Figure 9.2 Network capacity increase [Source: Neul]

But also, as shown in Figure 9.1, smaller cells have higher throughput. The net combination of these two effects is shown in the solid 'capacity gain' line in Figure 9.2 where it can be seen that a system with a 1 km cell radius has a capacity approaching 160-fold greater than one with a 5 km radius. Note, this does assume there are sufficient frequencies available that adjacent cells are rarely using the same frequency and so inter-cell interference is not problematic. If this is not the case and cells frequently use the same frequencies then the gains will be lower than shown here.

Hence, if capacity becomes an issue in a Weightless network, it can be addressed by smaller cells, albeit at a cost of increased infrastructure.

For likely cell sizes, it can be seen that cell capacity levels are likely to be below 1 Mbits/s. This helps size the backhaul requirements for Weightless cells. Most cells need only around 1 Mbits/s symmetrical backhaul but without contention since traffic generation may be near continuous. This implies that uncontended DSL would generally be adequate.

9.2 Factors that impact network capacity

Single cells do not occur in isolation and there are many other factors that influence the capacity of a network, almost all in a negative manner. These include:

- The percentage of terminals that are indoors.
- The amount of interference from TV transmitters.
- The amount of interference from other white space users.
- The number of frequencies available which impacts on the ability to use different frequencies in neighbouring cells and hence the self-interference.
- The quality of antenna on the terminals.
- The degree of 'throttling' if any that is applied to terminals with poor signal levels.

This section looks at the impact of some of these factors on the network capacity listed above.

Indoor terminals

Terminals that are located indoors will experience a weaker signal than those located outdoors. The difference is often called the 'indoor penetration loss'. It is very difficult to characterise because of the huge variation in buildings. At these frequencies typical models assume around a 15 dB mean loss with a standard deviation with $\sigma = 4$ dB. (With a Normal distribution there is a small probability of 3-σ events, hence this implies a

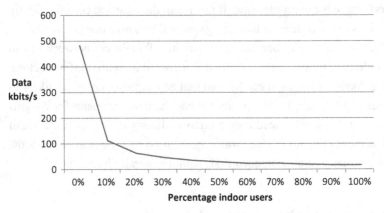

Figure 9.3 Variation in capacity with percentage indoor subscribers [Source: Neul]

range of values approximately from 3dB to 27dB.) The impact of having many indoor terminals is the same as extending the cell range or clustering terminals more towards the edge of the cell. Figure 9.3 shows how the average throughput for a particular cell radius falls as the percentage of indoor subscribers grows.

There is an enormous impact, reducing the cell capacity by over an order of magnitude when indoor subscribers grow from 0% to over 50%. If this is an issue the only effective solution is to reduce the cell size (or to find some way to place terminals outdoors or near windows).

Interference from TV transmitters

As explained in previous chapters, it can be expected that there will be interference from distant TV transmitters on many frequencies. Figure 9.4 shows how the data rate can vary for one particular case when a null is steered towards the interfering transmitter.

This suggests that nulls of 10 dB or more will be important within the network in maintaining channel quality. Beyond around 10 dB the gains become much smaller. Practical null steering systems can probably achieve between 10 dB and 20 dB.

Throttling

The cell capacity is typically dominated by the few worst users in the cell. This is because their data rates might only be around 1/100th of other

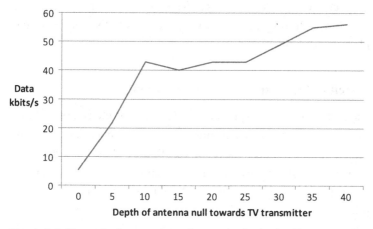

Figure 9.4 Change in data rate depending on the depth of null [Source: Neul]

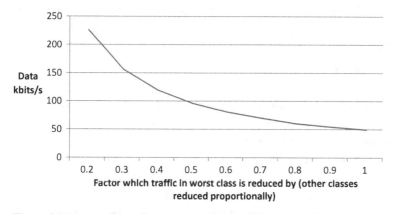

Figure 9.5 Impact of throttling poor-quality users [Source: Neul]

users so they require 100-fold more resource. Hence 10 poorly located users could take up as much transmission time as 1000 well-located users or 10 000 users with very good signal. Figure 9.5 shows the impact of 'throttling' or reducing the throughput provided to those users with the lowest signal levels.

Reducing the data rate by half would double the cell throughput. Such throttling might not always be possible but for some applications there may be some acceptable degradation – for example some smart meters might only send readings half as frequently as others.

The impact of other issues such as reducing number of frequencies are much more difficult to model but will be discussed in the context of the UK-wide network below.

9.3 Capacity of a sample UK-wide network

The figures presented in the previous section hint at the relative differences certain changes can make but do not give much guidance on the absolute levels. Indeed, many of the figures above use different assumptions to illustrate particular points, with the results that capacity levels differ radically across them. But which is the 'correct' set of assumptions?

To provide some further guidance on what might be achieved with a typical Weightless deployment, this section presents the results of a model of a realistic network in the UK. Indeed, this is a highly detailed model, taking many hours of processor time to run, which aims to be as realistic as is possible outside of actual deployment.

The model:

- Takes around 6000 real base station locations in the UK based upon a set of known available locations which are distributed approximately in line with population density.
- For each of these real base stations, the model consults a white space database to determine which frequencies the database would return (this has already been introduced in Chapter 3).
- It then determines the residual TV signal on each of these frequencies based on a prediction of TV coverage on every channel across the entire UK.
- It places a terminal in or on every building in the UK of which there are 24 million based on the UK postcode information providing building location.
- It then runs a central planning algorithm to generate frequency hopping sequences for each base station and considers the interference when neighbours or near-neighbours have to use the same frequencies.

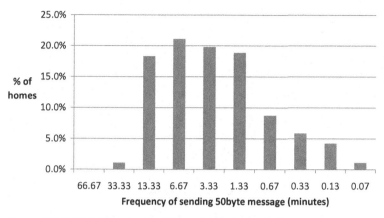

Figure 9.6 Polling frequency for UK model [Source: Neul]

- It then simulates service to every terminal across the network and monitors the data rate taking into account all the interference (both from TVs and self-interference).
- It makes certain assumptions about Weightless such as the antenna nulling achieved, transmit power levels, etc., which are thought to be representative of real systems.

While it would be possible to present the output from the model in terms of average capacity per cell, it is not clear that this is very helpful. In most machine applications the data rate is broadly irrelevant, what matters is that the machine can deliver its payload in a certain time period. For example, for smart meter applications the average message from the meter is around 30 bytes and this might need to be delivered every 30 minutes. That equates to a data rate of 1 byte/minute! Average cell data rates are of little use without further information on the number of terminals in the cell, their distribution, etc. Instead, a better metric is 'polling time' for each terminal in the network. For an assumption of a given message size it is possible to determine how frequently each terminal in the cell, and the entire network, could send such a message.

The results of the UK-wide modelling presented in polling time are shown in Figure 9.6.

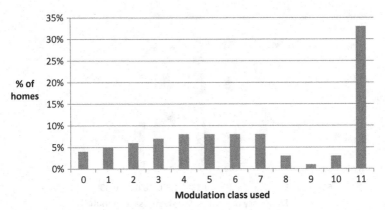

Figure 9.7 Distribution of terminals across modulation classes [Source: Neul]

This is based on:

- 6,000 base stations covering the UK each with a single channel and no sectorisation.
- 24 million terminals sending messages of 50 bytes.

It shows that the average polling time is around 5 minutes per terminal. That is, on average, a reading could be taken from each terminal every 5 minutes. If the requirement were for 30-minute meter readings on average then the network could accommodate 6 times as many devices – around 3 per person in the UK. Obviously, if meter readings could be stretched to e.g. 2 hours, this would increase by a factor of e.g. 4. However, there are some terminals that do have a polling time of around 30 minutes – about 1% in this case. These are typically located in cells which are heavily loaded, either because they contain a large number of terminals or more typically because they cover a large area and some terminals are 5 km or further away from the base station and hence use a very large fraction of the available resource. So if a 30-minute polling time had to be guaranteed, rather than delivered as an average, then this network would only just be sufficient. Of course, re-engineering the 1% of sites with congestion, as discussed further below, would address this problem.

Other instructive information can be gained from the simulation. Figure 9.7 shows the distribution of terminals across the modulation classes.

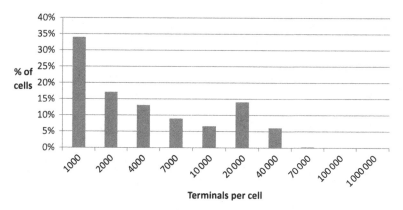

Figure 9.8 Distribution of terminals per cell [Source: Neul]

A large percentage is in the highest modulation class. This suggests that if there were even higher classes (e.g. based on 64-QAM rather than 16-QAM) there would be terminals that could take advantage of this. But with the capacity dominated by the furthest-out terminals this would make less than 1% change to the capacity at the expense of increased terminal complexity. Otherwise, the terminals are relatively evenly distributed across the classes, notable with some 4% in the lowest class. The apparent 'gap' around class 9 is likely a result of the propagation model used which adopts free space path loss for terminals close to the base station then transitions to Hata further out. The transition point tends to result in a fairly abrupt increase in path loss (which is often reflected in reality when the base station disappears from sight) with the result few terminals have a path loss value in this range and hence the corresponding modulation classes are less used. In practice, the real world is more chaotic likely resulting in a more even spread.

The distribution of terminals per cell is shown in Figure 9.8.

This shows that many cells have relatively few terminals. These are typically rural cells. The distribution of cell range is shown in Figure 9.9.

This shows a large number of cells with range 4–5 km. This is towards the upper end of the supportable range, especially for indoor terminals.

One issue raised earlier was the availability of white space. The impact of fewer white space channels depends on how many there are initially.

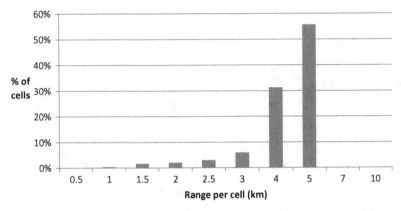

Figure 9.9 Distribution of cell ranges [Source: Neul]

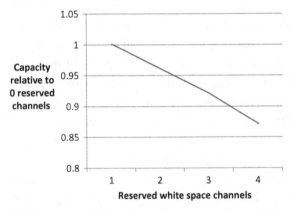

Figure 9.10 Impact of removing white space channels [Source: Neul]

Running our realistic simulation and then removing a certain number of channels nationwide leads to the result shown in Figure 9.10.

This shows a relatively graceful degradation in capacity, with the removal of one or two channels having relatively minimal impact (compared to e.g. increasing the percentage of indoor terminals).

Overall, then, this deployment is acceptable in that it would deliver a network able to meet smart meter requirements for all buildings within the UK. It does not have large amounts of spare capacity and in this arrangement could not meet the requirements of 10 devices per person if they all had 30-minute polling requirements (but it could if they had 2-hour average polling requirements). However, the analysis of performance shows that with changes to only a few per cent of the base stations the network performance could be markedly improved. How this might be done is discussed in more detail in the next section.

9.4 Increasing capacity

Cellular networks have shown how it is possible to increase the capacity over time. To do this they make use of techniques such as:

- sectorisation (taking a circular cell and turning it into a 3-sector arrangement)
- adding more carriers
- cell splitting
- using underlay cells
- traffic off-load.

Most of these techniques are available to Weightless. The analysis above shows how the capacity issues in Weightless are generally related to subscribers far from the base station. Approaches that target this particular problem tend to work more effectively.

Sectorisation is helpful. This can extend the signal strength through the use of directional antennas within the sector. Higher signal levels mean that lower spreading factors can be used, significantly increasing capacity. If mast space is available this is probably the lowest-cost option to improving capacity within a cell.

Conversely, adding more carriers is less effective. It does not address the issue of the far-out devices and hence only increases capacity in line with the percentage increase in carriers (so moving from one carrier to two would double capacity). Further, there are situations where there is insufficient white space spectrum to support multiple carriers in neighbouring cells and so the self-interference would increase, reducing the gains below that initially expected. Only in dense urban cells and where there is ample white space might this be useful.

Cell splitting can be extremely effective. As demonstrated earlier, capacity gains are enormous due to the combination of more cells and higher throughput per cell. Taking cells where many terminals are 5 km or more from the base station and inserting cells in between the existing cells might reduce the effective cell radius from 5 km to 2.5 km, with capacity gains of perhaps 15-fold. However, often the largest cells are in rural areas where finding new sites with power and backhaul might be problematic. (In passing, this is the opposite of cellular systems where cell splitting is employed in urban areas. This is because cellular systems do not have such a wide range of modulation and coding schemes and so the users with poor-quality signal are less dominant, and also throttled by many operators.)

Using underlay cells or off-loading is probably not relevant for Weightless. Both techniques target traffic hot-spots whereas it is the remote subscribers that are the issue for Weightless.

Determining the impact of these techniques on the capacity of the modelled network above would require detailed geographical analysis which is beyond the scope of the current work. It seems likely that a relatively small increase in cell count of perhaps 10% could readily result in a 2–5-fold increase in capacity.

9.5 Network cost and business case

Based on the above network, approximate business cases for a nationwide Weightless deployment can be created. To do this, the assumption has been made that Weightless is deployed by an organisation that is either an existing operator or has access to the necessary infrastructure. For example, this could be a cellular operator, a fixed line operator or a mast owner. Further it is assumed that Weightless can be deployed on existing masts and that no additional backhaul is required although there is some cost-share of the current backhaul.

Developing the business case at this stage requires many assumptions which may prove incorrect around factors like base station costs so the following must be treated as highly approximate at this stage.

Base station costs might be of the order of $10 k each including antennas. Allowing a further $5 k for installation and assuming 6000 base stations the hardware investment is about $90 m. There are no spectrum costs and no up-front core network costs as this is a cloud-hosted service. Annual network costs might include around $6 m for backhaul, $9 m for maintenance and perhaps $10 m core network and support fees. Amortising the network deployment over 15 years (and ignoring discount factors) leads to around $34 m per year. On top of this would be staff, premises, marketing, customer relationship management and more, perhaps a further $20 m per year. Overall costs might fall in the region $50 m per year.

Smart meter tenders have suggested annual subscription costs of around $10 per meter. This would amount to $240 m per year in the UK. If the network supported gas and water meters as well this would rise to around $720 m. Add in other machine applications and the revenue could easily exceed $1 bn/year. Of course, revenues will fall over time and competition will emerge. Nevertheless, it is clear that these estimates would have to be orders of magnitude in error for the business case not to be extremely profitable.

9.6 Comparison with mesh systems

There are two basic architectures for wireless systems – mesh or star topographies. Star topographies are much more common and consist of a central transmitter (a base station in cellular, a router in Wi-Fi) which acts as a controller to wireless devices and then routes their message back into the core network via a backhaul connection (leased line in cellular, DSL or similar in Wi-Fi). Mesh topographies dispense with the central transmitter; instead devices relay signals from one to another until they either find their destination or a 'sink point' where information can be routed back via a backhaul connection to the core network. In some respects the sink point in mesh is similar to the base station in a star system except that the sink point only communicates with the devices very close by (which are relaying signals from devices further out). Weightless uses a star topography.

Mesh solutions are conceptually elegant. They do not appear to need an infrastructure, they can re-route messages in the advent of link failure in a similar manner to the Internet and because each device only transmits to a neighbouring device then lower power levels appear possible. However, mesh systems have proven difficult to implement in practice. They suffer a number of general problems:

- Finding satisfactory routes to a node can be very difficult, especially if devices are moving requiring permanent recalculation. For this reason, practical mesh systems are often restricted to stationary devices.
- Starting the mesh is problematic. Until there is a high density of devices, there may be some devices that are out of range of all other devices. But typically coverage is required from the moment a device is installed. This can lead to a need to 'seed' the mesh with additional devices whose sole purpose is to fill in gaps.
- The need to relay messages can lead to much additional traffic and delay on the messages.

For these reasons, mesh systems have not had widespread deployment to date but have been suggested for some machine applications.

The key problem with mesh for machine communications is the difficulty of handling moving devices. This makes the provision of a complete machine communications solution hard. The applications mesh might be able to cover include energy, industrial and smart city-type solutions but the applications it may not be able to cover include transport, automotive, asset tracking and healthcare. It is difficult to estimate the size of each market but it seems likely that the static element might be 50% of the marketplace or less. This could result in higher cost per user, lower economies of scale and the likely emergence of a star network to handle the other applications (which might as well then handle the static applications too).

Even in static applications, mesh may not be optimal. Latency can be an issue where there are multiple hops, especially if each device uses sleep modes to conserve power and hence is only available for relaying for certain periods.

Broadcast is an important feature in machine communications. This might be used to send a software update to multiple devices or to send 'alert' messages to a complete class of device. In a star topography broadcast is simply achieved by transmitting a message as normal but instructing all devices of a certain class to listen for it. In a mesh, broadcast is more complex, involving a terminal listening to a message and then relaying it on to all of those nearby. The use of multiple re-transmissions is inefficient and ensuring that all devices get the message difficult. Not all will get the message at the same time which may be a problem in applications such as energy management.

Mesh systems cannot readily access white space spectrum. This is because white space requires a master–slave topography whereas mesh systems have multiple slaves. Each terminal would need to go through the geo-location process of determining where it is and then sending a request to a database, possibly using an alternative radio mechanism. Ways to cascade frequency assignments through the mesh could be envisaged but are highly complex and risk devices losing contact. This means mesh systems typically use spectrum at higher frequencies with less good propagation and often less bandwidth.

Mesh is generally considered when it appears that a star system will not work. It has been said that this is the case for machine communications because low-powered sensors running for years on batteries will have such limited range that unless an ultra-dense infrastructure is built they will be unable to communicate. But such an infrastructure would be overly expensive for machine solutions.

However, the advent of the white space spectrum changes this dynamic for two reasons. Firstly, the low frequencies of the UHF band provide excellent propagation, extending range much further than, e.g. the 2.4 GHz band. Indeed, range is about four times greater, resulting in a need for only around 1/20th as many base stations. Secondly, the wide bandwidth channels allow for the use of spreading which further extends the range, albeit at the expense of reducing data rates. In the TV bands channels are 6 MHz or 8 MHz wide enabling spreading by up to around 1000-fold where needed, adding some 30 dB link margin. Through this combination of low frequency and spreading, ranges of 10 km are feasible from battery-powered devices, overcoming what appeared to be a major problem for star architectures.

10 Application support

10.1 Introduction

Dr Antony Rix[1]

Weightless could be suitable for a wide variety of applications, from smart metering to healthcare and even asset tracking, as noted in Section 1.3. In this chapter we consider which application requirements fit well with Weightless, explore what is required to build a system that integrates with Weightless, and then look in more detail at example applications in automotive, energy, healthcare and consumer markets. This chapter is not a complete guide on building end-to-end services – substantial organisations exist with this as their main focus – but is intended to provide an informative introduction to some important considerations.

Although Weightless is intended to be reasonably application-agnostic, it is important to understand that design decisions made in defining the standard, and the unlicensed nature of white space spectrum, place important limits on which applications will work best with a Weightless service. For example, while half-hourly readings from a utility meter could easily be sent over the Weightless system, video streaming would likely overload it.

Conversely, the requirements of some applications and business models do need to be addressed within the system, even potentially in the physical layer. Generally these requirements will be common to several markets, so the solutions are of quite generic benefit. Security is one of these requirements, and something which cellular networks address in detail. A topical example of why this is important could be the following.

---

[1] Dr Antony Rix is a Senior Consultant at The Technology Partnership plc (TTP), where he leads projects developing wireless products and services. Prior to this he co-founded a successful start-up. He has an MEng degree with distinction from Cambridge University and a PhD from Edinburgh University, both in Engineering. Antony co-chairs the Cambridge Wireless Connected Devices SIG, and is a member of the American Telemedicine Association, Audio Engineering Society and IET.

Consider a Weightless-connected diabetes monitor used by a celebrity. Machine IDs like MAC addresses are often assigned to manufacturers in ranges, each range to be used by a particular product to provide traceability, and thus identify not only the vendor but even the type of device. If the machine ID is sent unencrypted over the air, it might be possible for an attacker with a channel monitor to deduce that the celebrity has diabetes, track the celebrity's movements, or even guess when the device is used. Appropriate security precautions in the system can make this difficult or infeasible.

Many applications also need much more than just periodic, secure communication of data. Devices need manufacture, configuration, enabling/disabling, software updates and billing. Business relationships need flexibility to support several business models, multiple network providers and to work in different countries. Many of the companies who can benefit most from Weightless are those whose current business model is to sell devices at arm's-length to end users, and while Weightless could allow them to build a customer relationship and gain both differentiation and service revenue, these companies have often never done anything like this before.

This chapter therefore looks at what makes an application well-suited to Weightless, and then evaluates the requirements for an end-to-end service platform that will be common to many applications. Examples of what Weightless could enable in several markets are then discussed.

10.1.1 Matching applications to Weightless

Communications networks can be categorised according to several well-understood and measurable criteria that have a substantial impact on how applications work. This section highlights a number of these criteria to allow the reader to consider whether a particular application would work well with Weightless.

Latency, maximum latency. These terms are used to describe the typical, and maximum, times between a packet of data being offered to the network for transmission and when it is received and acknowledged. The maximum latency may be most important for time-critical applications like raising alarms, while the typical or average latency is often considered if a person is in the loop. For voice communications, latency is ideally only a few milliseconds to avoid echo and conversational impairments.

Many interactive Internet applications have latency on the order of a second. A postal delivery service, in contrast, might have maximum latency of 48 hours or more. Weightless systems are expected to have latency of several seconds although this will depend on system configuration and load (see Section 10.9).

Message size, message rate, data rate. Applications often send information in messages of known, regular message size – many applications need only a few numerical values to communicate readings and status, with message size of a few bytes. A software update for a type of device could easily be several megabytes in size. The message rate determines how often messages are to be sent, and a second consideration is whether messages occur on a regular schedule or may be generated by some other, unpredictable process. The data rate is the product of the message size and rate.

Weightless is well suited to small messages and message rates from around one per minute upwards, especially in the uplink from devices to the network. In the downlink it can achieve much higher data rates permitting occasional large transfers like software updates (see Section 7.11).

Coverage. Cellular users are well aware that wireless networks do not provide coverage everywhere – UK 2G networks provided 96% of postcodes with at least 90% probability of (outdoor) coverage in Q2 2011 according to Ofcom [1]. Yet a device that relies on Weightless for a key aspect of its function absolutely requires coverage where it is going to be used, which will often be deep inside buildings.

The coverage of a Weightless network will depend to a large extent on spectrum being available and the density of base stations, but many of the design decisions behind Weightless, such as the use of UHF TV frequencies, support very long range that means coverage is likely to be high, even in buildings. Coverage for a possible UK Weightless network is discussed in Section 9.3.

Antenna size and location. These aspects of the radio can have a surprisingly large impact on the performance of an application, and developers new to wireless often underestimate the importance and the difficulty of integrating the radio antenna to achieve high performance, and product cost and design objectives like size and case material are often in conflict with radio requirements. Even major smart-phone models have reportedly shown performance 5–10 dB worse than expected. A reduction of 1 dB in antenna performance has a direct impact on network cost, requiring 14% more sites to achieve the same level of coverage [2] – so a 10 dB performance drop requires about 4 times the number of base stations.

While the long range of Weightless can partially compensate for a poor antenna, application developers must understand how fundamentally important antenna performance is, or risk the potential of high product return rates. Weightless operators need to consider radio performance as it has a direct effect on the coverage and capacity of the network. This is especially important as the relatively low frequencies used for UHF mean that the antenna needs to be larger than the corresponding cellular device to achieve high efficiency, while the very wide frequency range makes it difficult and potentially expensive to achieve adequate efficiency in all available channels.

Some of the fundamental points of antenna design are to keep the antenna away from materials like the human body or a metal case that could block or absorb the radio, to orient the antenna so that radio energy is transmitted in the required directions, and to choose an antenna structure that offers high efficiency over the necessary frequency range and a good radiation pattern. However, highly specialised techniques are needed to achieve maximum performance. The details in Figure 10.1 illustrate the difference in radio antenna gain between medium size ($50 \times 50 \times 2$ mm) a low-cost antenna built as a printed circuit board track, and a low-cost tuneable antenna of smaller footprint ($8 \times 50 \times 8$ mm) designed by a specialist.

Availability. This measures the proportion of time, typically over a year, for which the system is functioning normally and permitting expected use. In any communications system, there is always the

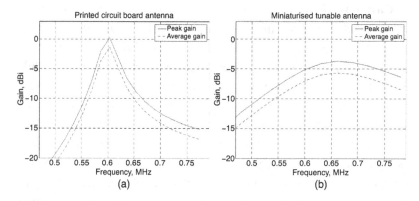

Figure 10.1 Comparing antenna gain between simple and high-performance designs for a small portable TV-band device [Source: TTP]

potential for equipment failures and intentional attacks, which should be considered throughout a system's lifecycle. A wireless system is also subject to interference from sources ranging from other base stations and users on the same network, to other competing networks, and other types of equipment that may share the same, or close, radio spectrum. This is a key difference between White Space and cellular systems because the latter are typically exclusively licensed, while potentially anyone can operate a White Space system providing it conforms to rules such as the access database. The large amount of TV spectrum available in most locations can help Weightless achieve high availability, but no wireless system can be expected to have 100% availability.

Quality of service. Even if coverage exists and the network is available, many communications technologies do not guarantee delivery. There are several ways to define quality of service, and a relevant metric for Weightless would be the proportion of messages sent in each direction that are correctly delivered to their destination and successfully acknowledged, where the sending device believes it has both coverage and network availability. Through the use of retries, it is expected again that this will be high, but the potential for interference means quality of service will also be lower than 100%.

Cost. A key factor in providing communications for many applications is how much device cost the radio, antenna, SIM card and other elements add to the product. Even in large volumes, as of 2011 GSM technology still costs on the order of $10 per device when fully integrated, while Weightless targets a device cost that is around one order of magnitude lower. The service cost is also an important amount to consider, as the operator of a Weightless system must be able to pay back the costs to build and operate the network of base stations and the supporting systems, and is likely to do so by charging for usage such as data transfers. While comparable data about Weightless is not yet available, figures for very low usage GSM machine plans indicate that annual service costs are of similar magnitude to the device cost in the most competitive markets, but there is considerable variation. Clearly the application must add value in some way that can sustain these costs.

Other considerations. The list above is not exhaustive and particular applications have many different needs. These could include battery life, security, drop protection, water resistance and so on. Like any communications technology, Weightless will have certain requirements and properties in these areas that can influence the potential of an application.

10.1.2 Building an end-to-end service using Weightless

Let us now consider the steps to be taken to put into production an application that is well suited to using Weightless, based around a hypothetical example of a home weather station. This provides a convenient overview of the key process elements and the main system components.

Business case. The weather station manufacturer is developing the next generation product, and has commissioned a feasibility study on how to enable the product for Weightless and provide a networked service. Market research has indicated that the Weightless-enabled product could be sold for $20 extra including the additional device cost and the first year's service cost bundled in, with subsequent years' service costs being an acceptable $10/year. The business model indicates that, with amortisation of the estimated development cost, appropriate supply and distribution chain mark-ups and taxes including the Weightless module, antenna and network fees, this would just break even in the first year and would then yield a gross profit of $5/year for every user that renews the service plan. The business thus decides to go ahead with the project and engages a development team to realise the product.

Development. The prototype already includes a microprocessor and a communications interface such as RS-232 that could be used to send readings to the Weightless module, and a power supply that can be used to drive it. The development team obtain Weightless development boards from an electronics distributor that come bundled with access to a Weightless operator's development service platform that is functionally identical to the operator's production service platform. The development team adds functions to the microprocessor to communicate weather data periodically to the Weightless module and to read and display the service availability status. At the same time, the development team builds a web platform that allows users to set up accounts, provide address and billing details, and securely enter the IDs of products they have bought to enable the service. The web platform is integrated with new back-office systems to securely process users' personal data, allow telephone support, and to store and serve up the weather data for display via the web platform on users' PCs and phones, and the prototype system is tested.

Commissioning. The development team integrates the Weightless module into the prototype product, updating the circuit boards and case-works, to produce a pre-production prototype that is also tested. This is passed through lab tests, regulatory, Weightless, operator and other necessary approvals and the design is transferred to the manufacturing facility. A production test process is put in place to ensure that the devices are fully working with a Weightless network before leaving the factory. The back-office systems are scaled up to meet the expected traffic and tested and audited exhaustively. The business also obtains the approvals needed to market the product and service and process customers' data and payments, and finalises the necessary contracts such as service provision. A first production run is ordered using production Weightless modules and checked both in the development team's lab and with employees and friendly users, and following satisfactory tests the product is now ready to ship to distributors and end customers.

10.1.3 The profile concept

If Weightless was to be developed in a way that is independent of applications, it might be possible to ignore applications in the standard and in this book. Weightless modules could present a uniform interface to all applications, providing low-level functions such as a method of monitoring the radio link and sending or receiving messages of general format. Indeed, it is quite likely that some such interfaces will be created for the internal use of the module or to allow the greatest freedom to developers who may prefer to have proprietary software of their own running on top of the standard system.

However, as noted above, it is difficult to consider applications entirely in isolation from the radio, and certain applications would appear to have a strong fit with Weightless. It is also common for user groups or developers in some markets, particularly those that are highly regulated, to seek to collaborate. This may be motivated by a desire to avoid the repeated investment associated with creating incompatible yet similar solutions, to drive multi-sourcing to the extent that different vendors' equipment becomes interchangeable and interoperable, or to foster development of reusable components or devices by third parties.

This leads to the idea of the profile, which is already well established in some wireless standards such as Bluetooth [3] and Zigbee [4]. A profile typically defines a set of standardised, mandatory behaviours and interfaces that a particular application must implement. These may be rigorously and independently tested, and often are intended to provide interoperability where devices from different vendors may need to work together.

A profile can also codify the characteristics, requirements and features of the application, though it is not the only way to do so. In the case of Weightless, profiles could define the type, size and frequency of message sending, or the extent to which a device can be mobile. This knowledge can have a substantial impact on how the network interacts with the device and this information can be of benefit to both.

A good example is Bluetooth's Headset Profile (HSP) [5], which allows audio headsets from any manufacture to work with audio gateway devices, typically mobile phones or PCs, to allow voice calls to be placed, received or transferred to the headset. By defining this profile, the founders of Bluetooth made it possible for an ecosystem of competing manufacturers to evolve that would provide headsets that, at least in theory, could be used with any of their products. Traditional Bluetooth has developed over 20 profiles for different types of device or use-case [6].

It should also be noted that the ultimate goal of a profile – to define a completely vendor-independent, backward and forward-compatible interface – is extremely difficult to realise in practice, especially if the market becomes as fragmented and the protocol complex as is the case with headsets. An Internet search rapidly returns large numbers of reports, including vendor support websites, illustrating this point.[2]

While there are many potential advantages to defining profiles, there are also disadvantages such as the time and investment required in standardisation and the erosion of competitive advantage. Weightless also differs in critical respects from Bluetooth and Zigbee. It is not yet clear whether or when the development of profiles will be carried out for Weightless; what is certain is that the developers of Weightless have the opportunity to learn from the experience of these earlier technologies and develop a system with generic interfaces that meet the requirements of many applications in an open and scalable manner.

10.2 The connected car: applications and challenges

Iain Davidson[3] *and Andrew Birnie*[4], *Freescale*

The phrase 'connected car' now means everything from live GPS, toll payment, telematics and pay-as-you-drive, to streaming video across

---

[2] For example, the search term 'Bluetooth headset incompatible' returned about 274 000 hits on Google in December 2011.

[3] Iain Davidson (Freescale Semiconductor UK Ltd.). A graduate of University of Strathclyde in Glasgow, Iain has been 18 years in the embedded systems industry where he has held various engineering, systems and management positions. Iain now works in business development, promoting Freescale's networking technology for Machine-to-Machine (M2M), Wireless Sensor Networks (WSN) and Internet-of-Things (IoT).

[4] Andrew Birnie (Freescale Semiconductor UK Ltd). After graduating from the University of Glasgow, Andy had various roles in microcontroller product and technology development, but is currently Systems Engineering Manager for automotive body electronics and driver information systems within Freescale. In this role Andy is responsible for understanding market trends and customer demands, to define the next generation of microcontrollers.

Figure 10.2 Connected cars in the smart city. [Source: markarma]

the Internet, hands-free mobile telephone, eCall[5], remote diagnostics and car-to-car communications for collision avoidance. The number of connections and applications of those connections is potentially vast.

Several of these connected car applications have been trialled or demonstrated around the world and it is fair to say that some are more realisable than others, some are already quite mature, some have very clear standalone business models and others might well fail for lack of true commercial value. Frustratingly, those applications that will be slowest to emerge into production vehicles will be safety related, where the question of 'who pays?' is particularly important. The scope of applications for the connected car is only limited by our imagination.

In the smart city of the future, illustrated in Figure 10.2, everything will be connected; our buildings and homes, people – their bodies, their smart-phones and tablets; our energy generation and supply

---

[5] eCall is a project intended to bring rapid assistance to motorists involved in a collision anywhere in the European Union. http://ec.europa.eu/information_society/doc/factsheets/049-ecall-en.pdf.

infrastructure; our factories and our industrial and civic infrastructure; transport infrastructure, the traffic signals as well as trains, buses and of course cars . . . connected cars.

This transformation we are working towards will see all these machines connected to the Cloud and part of the Internet of Things. While our fixed and mobile networks are inherently secure, the endpoints are not (always) and just look at the absolutely vast number of connected nodes we will have – often in fairly accessible places. As a result it is important to build trust and security into our networks and nodes – a vital point for machine networks and in particular the connected car. We will touch more on trust later, for now, let us remind ourselves why it is worth the effort.

The overall benefits of connected car are widespread and illustrated briefly in Figure 10.3. There can be benefits for the driver as well as for the automotive industry and the services around it and equally there can be benefits to the economy and to society as a whole. Some applications could benefit all.

These points are illustrated by the following examples:

- The World Health Organisation reports that over 1.2 million road deaths occur globally each year. Wider deployment of existing safety systems would help. Connected car applications would further enhance safety on our roads, for drivers, passengers and other road users alike.
- In large cities, as much as 50% of traffic congestion was caused by drivers cruising around in search of a cheaper parking space [7]. It is tempting to speculate that this will be compounded by electric vehicles looking for a charging station.
- One might assume that all cars should be connected to see benefits but research in Germany recently suggested that only 0.5% of cars thus connected can make a difference to traffic congestion [8].

The benefits themselves can often be multi-dimensional. For example, consider a connected car application built on detailed telemetry data gathered or uploaded during each journey made. This could be exploited in several ways;

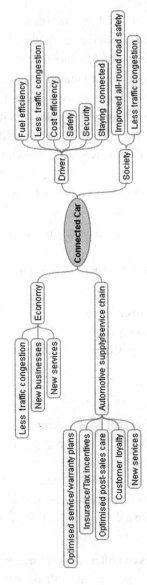

Figure 10.3 Connected Car benefits

Table 10.1 *Connected car applications*

Connected Car	Connected driver/passengers
Core Auto Systems	Driving related
• Remote diagnostics connection	• Traffic information connection
• Engine control unit (ECU) software upgrade	• Mobile navigation connection
	• Electronic toll connection
• Sale of vehicle software options	• Road usage fee/insurance connection
• OEM/Dealer link	
• Electric vehicle battery status	• Driving style/Greener driving
Safety & Security	Safety & Security
• Car-to-car: driver assist systems	• eCall connection
• Car-to-roadway: collision avoidance	• roadside assistance link
Car infotainment Systems	Infotainment related
• Remote system control	• Voice communication link
• Map update connection	• Information access connection
• Stolen vehicle tracking connection	• Entertainment download link

1. The driver can learn about their driving behaviour and its impact on fuel efficiency. By driving greener, the driver will probably be a safer driver.
2. The safer driver will be less of an insurance risk and might benefit from cheaper insurance in future.
3. The greener driver will consume less fuel and produce less CO_2. An electric car will consume less electrical energy.
4. The car will probably last longer, at least require servicing less frequently

Similarly, effective traffic management will benefit individual drivers, and by extension all businesses which use roads logistically will be better off as will society as a whole.

Let us look at some specific connected car applications (Table 10.1) and consider those which benefit the driver (connected driver) or the car original equipment manufacturer (OEM) or car dealership but we

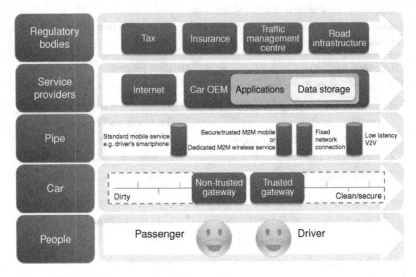

Figure 10.4 Multi-layer connected car model

must not forget the positive effects these can have on society and on the economy.

It is true that fleet management and sat-nav technology exists today and provides point solutions for several of these applications but here we argue that an integrated approach is key for mass adoption and for delivering on the multi-dimensional benefits shown in Figure 10.3.

That said, there remains a question mark over whether a single architecture can deliver this. Perhaps obvious already is the need for a multi-layer and multi-player approach but there are also different architectural approaches each with their own advantages and disadvantages.

We will explore the merits of the in-car gateway connectivity and the different 'pipes' later. We will first build on the connected car applications by describing the bandwidth and trust levels required for some of the applications in each category before mapping them to the multi-layer model shown in Figure 10.4.

When you weigh up the different usage, bandwidth, trust parameters, there are four distinct categories emerging;

Table 10.2 *Connected car application parameters*

Example use-case	Benefits	Usage	Bandwidth	Trust level
Remote diagnostics connection	Driver, OEM	24/7 or @home	Medium	High
ECU software upgrade	OEM	@home	High	Highest
Sale of vehicle software options	OEM, driver	@home	High	High
Electric vehicle battery status	Driver	24/7	Low	High
Electronic toll connection	Driver	24/7 or @home	Low	High
Driving style	Driver, Society	24/7 or @home	Medium	High
Road usage fee/insurance	Driver	24/7	Medium	High
Map update	Passenger	24/7	Medium	Basic Internet security
Entertainment download	Passenger	24/7 or @home	High	Basic Internet security
Traffic information	Driver	24/7	Medium	Basic Internet security
Roadside assistance	Driver	24/7	Low	Basic Internet security
Collision avoidance	Driver, society	24/7	Low	High
Driver assist systems	Driver, society	24/7	Medium	High

Notes: 24/7 indicates a real-time or pseudo-real-time connection which is always on whether the car is in transit or stationary.
@home indicates wireless communication at dedicated locations, e.g. the home or office.

- OEM Software
- Multimedia/Infotainment
- Safety
- and a catch-all category we will call 'telemetry'

OEM software use-cases

The car just like your smart-phone or tablet and so many other devices we use today is software configurable. The modern car is quite probably the most technologically advanced item any of us own. Higher-range cars will have 50–70 microcontrollers within the vehicle controlling functions like anti-lock braking systems (ABS), engine control, air conditioning, dash-board and much more. From a software point of view, routine upgrades would be performed when you have the car serviced once or twice a year but it would be far more convenient to have this performed remotely over a wireless connection – your car gets the update sooner without a trip to the garage.

While it would be quite possible to perform these software downloads to a moving car (with the software image buffered and only applied when safe) it hardly seems worth any risk, since waiting until the car is stationary and connected to a fixed DSL or Wi-Fi is good enough. You still get the benefit, a benefit which is pretty clear and robust. The actual download would be simpler with a high bandwidth connection but a trusted/authenticated connection and download process is the key here.

Multimedia and infotainment (for driver and passengers)

Maps, email, video streaming, web-browsing, traffic information, etc. all have proven value and we do much of this in the car today via a smart-phone or tablet. Having a more integrated display and interface would certainly be more convenient – and in the case of the driver be moderately safer – but in a machine communications context these applications do not fit – they are smart-phone or tablet applications and can be best supported by them, a docking point is probably all that is needed.

Safety

Car-to-car and car-to-infrastructure communication can bring untold value to road safety. Often termed vehicle-to-vehicle (V2V) and vehicle-to-infrastructure (V2I) the characteristics are different from the other application categories in the sense that the connection is shorter range and requires low latency in the millisecond range. These use-cases involve short-range communications from car–car and car–roadside infrastructure to pass on localised information, and there might even be a case for V2I2V. Examples include:

- traffic lights will change in 3 seconds;
- an emergency vehicle is approaching;
- there is stationary hazard in front;
- there is an unseen vehicle approaching (a motor-cyclist perhaps);
- not safe to overtake (perhaps V2I2V).

The response time of current mobile networks is almost certainly insufficient to support these scenarios effectively and so a dedicated alternative called Dedicated Short Range Communications (DSRC [9]) is emerging which would likely be based on an 802.11p[6] standard, an extension to Wi-Fi. Despite the clear benefit, the lack of payment model and the uniqueness or solitary purpose of this solution may work against it. Fourth generation cellular (long term evolution – LTE) could help reconsolidate. More likely is that competing technologies based on radar and/or vision-based systems [10] will succeed by solving some (in reality just a few) of the use-cases which DSRC could address.

Telemetry (location based data)

We termed this fourth category a catch-all but that does not mean it is unimportant. In fact the opposite is true – telemetry-based use-cases or more accurately telemetry data can help launch some high-value connected car applications and raises real prospects for new innovative services to be built around it. Generally speaking, this category requires low-medium bandwidth, higher trust levels and pseudo-real-time performance. Within this category are applications such as:

---

[6] IEEE 802.11p is an approved amendment to the IEEE 802.11 (Wi-Fi) standard to add wireless access in vehicular environments (WAVE).

- traffic management/congestion reduction – just 0.5% of cars can help
- road conditions – a fusion of ABS activation, an accelerometer, temperature and GPS data
- tracking
- toll charging
- fuel/charge levels – time/distance to nearest top-up
- remote diagnostics – owner, recovery services, dealership, OEM
- remote maintenance – dealership, OEM
- driving style – acceleration, breaking, speed, fuel consumption, fuel efficiency
- traffic information – upload route/journey information to a Traffic Management Centre

We elaborated on the driving style use-case earlier which we now expand further.

The same telemetry data which could be used by the driver to be a safer, greener driver could be used in the insurance process to reward those who are more responsible and it could be used to optimise emission-based taxation to more accurately reward greener drivers. Additionally it could be used to assess traffic flow and form part of a traffic management strategy built on real-time telemetry uploaded by even just a few cars.

On the diagnostics and maintenance side, we again see multiple benefits. A single remote diagnostic capability could serve the owner, the recovery services, the car dealership and OEM alike, perhaps with different levels of capability.

The owner could be allowed to check levels and pressure, to lock doors, prime or de-activate alarms, to pre-heat or pre-cool their car, whereas the other agents would have incrementally greater access. The prospect of super-accurate remote and pre-emptive maintenance has benefits for all. Car OEMs and dealerships can help plan maintenance which if done well can improve customer retention. They could optimise their own service and warranty processes (pre-ordering replacement parts for example) and build new services potentially passing cost gains or savings back to customers or using to fund connect car technology roll-out.

Car OEMs who have had to endure the pain (to the bank balance and to their image) of product recalls might be particularly keen to leverage connected car technology. Being able to monitor the performance and behaviour of problem components in the field may help them avoid or better manage such incidents.

Selling or freely opening some or all of this data could open up new unimagined services or apps . . . the possibilities are huge.

Now if we take the views expressed in the previous discussion and map those to the multi-layered model then some clarity starts to emerge – and in fact we are close to concluding that one solution, one connection does not fit all the application use-cases. Before we conclude let us look a bit closer.

Until now we have only alluded to some critical factors such as trust and system architectures but as you will see, these have a significant bearing on future outcomes. As with many other connected machine applications the connected car requires trust to be built in from the ground up. Trust in the connected car and in the overall eco-system must deliver privacy where it is needed to protect the integrity of new services created around the telemetry data or the connected car itself.

Trust will also maintain safety. In a connected car, the connection represents a new threat dimension which if compromised will be a potential hazard for the car and more significantly could be a danger to driver, passengers, other road-users and pedestrians alike. Researchers at the Centre for Automotive Embedded Systems Security [11] starkly illustrated this point in 2010 when they demonstrated a lack of resilience to malicious attacks on the electronic control units of a car. While they succeeded in disengaging the brakes in a moving car it has to be said they were privy to inside knowledge and that it would be hard for malicious hackers to reproduce their work. The point has been made however.

There are different dimensions to the trust question, each having their own characteristics:

- data on the move – over the network;
- data at rest – in the car or in the cloud;
- secure access – who has 'keys to the car'.

Existing hardware and software technology from Freescale and the broader networking industry is already solving these challenges. The trust toolkit includes network encryption and authentication – we will rarely think of it but cryptography algorithms like Kasumi[7] in 3G, IPsec[8] and VPN[9] through the packet core keeps our networks secure. While events of the last year or so have suggested that some cloud-based services and networked devices are more robust than others, technologies such as firewalls, anti-virus and intrusion detection can be effective. At a gateway or device level, access and security can be further enhanced by using the latest trust architecture technology available on Freescale microcontroller and microprocessor Systems on a Chip (SoCs). Features such as secure boot, run-time integrity, anti-theft and anti-tamper provide a platform for trusted computing.

So the data and access challenges are solved or solvable. A natural continuation of the secure access topic is to look at the car-to-cloud system architecture. A second influence on this however is arguably the single biggest challenge to a successful mass adoption of connected car – who pays?

When considering this question it is useful to first look back at Figure 10.5 where the wide range of connected car use-cases are characterised as (i) OEM Software, (ii) multimedia, (iii) safety and (iv) telemetry.

Keeping those in mind, now think back to the last new car you bought and the options you were probably presented with. You might be surprised to discover that the most popular options taken by new car buyers are in a fifth category – (v) cosmetic.

---

[7] KASUMI is a block cipher used in UMTS, GSM and GPRS mobile communications systems. 3G, 2G and 2.5G respectively.

[8] Internet Protocol Security (IPsec) is a protocol suite for securing Internet Protocol (IP) communications by authenticating and encrypting each IP packet of a communication session.

[9] A virtual private network (VPN) is a network that uses primarily public telecommunication infrastructure, such as the Internet, to provide remote offices or travelling users access to a central organisational network. VPNs typically require remote users of the network to be authenticated, and often secure data with encryption technologies to prevent disclosure of private information to unauthorised parties.

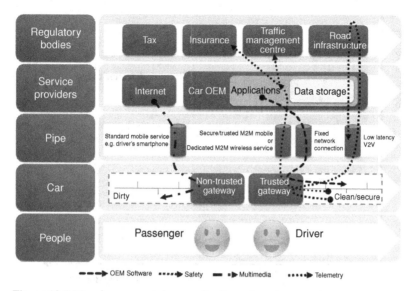

Figure 10.5 Mapping connected car application categories

The most popular options are things like paint colour and finish, leather seats and alloy wheels. We generally think we are all good drivers and do not need any additional safety systems.

If this mindset continues then we are faced with the prospect illustrated in Figure 10.6. Systems which help us and others be safer, which help traffic move more freely or which help us be better, safer, greener drivers will be slowly or minimally adopted. Conversely, systems which help us stay connected to work or entertain our passengers will be bought at a greater rate.

Regulation and/or legislation will clearly help incentivise but most important is for industry to work together across the different layers (Figure 10.4) to deliver a low-cost connection. Low cost both in terms of hardware and network transaction charges. The aim is for these costs to be so low that the services are effectively free.

Is this realistic? Let us look at different smart connected devices in everyday use today.

A smart-phone or tablet can be connected via Wi-Fi or 3G but we all use Wi-Fi when it is available because it is almost always cheaper. There

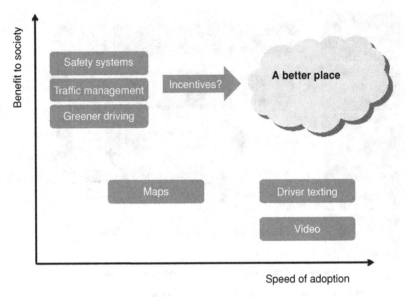

Figure 10.6 What sells?

is an important message here. For a connected car, Wi-Fi would be fine if the car is parked at home, at work or near a Wi-Fi hotspot but useless when it is on the move. 3G offers better mobility options for sure but then there is the cost. Few of us would be willing to pay for a second 3G subscription just for our car.

The eReader model on the other hand is a far better prospect. The same connectivity options are available – Wi-Fi or 3G – but the subscription cost is paid by the retailer or manufacturer. The hardware cost associated with the wireless connection is hidden within the overall device cost and any recurring service charges can be small.

So a successful model exists but how do the different connected car use-cases stack up against that model – is it workable? To assess, let us look back at Table 10.2 and reflect on the needs of the different use-cases or use-case categories.

- 3G gives us mobile broadband. Of the use-cases only the multimedia use-cases need mobile broadband – video streaming, web-browsing, live traffic info etc.

- There is a tranche of telemetry-based use-cases. Some are in use on the move but only require pseudo-real-time connectivity. For the others connecting at home is sufficient.
- The OEM software-related use-cases are almost certainly an at home scenario. The idea of a software upgrade mid-journey is at best emotionally problematic, at worst a safety risk.
- Finally there are some safety-related use-cases which could be argued are special cases. Vehicle-to-vehicle (V2V) and/or vehicle-to-infrastructure (V2I) for collision avoidance or driver assist require wireless connectivity with a rapid response time. DSRC is earmarked for this scenario however it would require dedicated single purpose radio and infrastructure.

At this point we can put aside two of these scenarios. Scenario (i) – OEM software use-cases – would benefit from broadband but not necessarily mobile broadband. Software downloads or uploads will be done at home. Scenario (iii) – safety using V2V/V2I is a special case requiring special single-purpose infrastructure.

This leaves us two main use-case categories – scenario (ii) – multimedia/infotainment and scenario (iv) – telemetry. As we can see from the mapping in Figure 10.5 these are two fundamentally different scenarios with quite different solutions.

- Multimedia/infotainment data comes from the Internet and does not rely on or impact any trusted in-car network. There is an argument therefore that the driver or passenger smart-phone is the best device to support these use-cases over 3G or indeed Wi-Fi. The smart-phone could easily be docked and connected to in-car displays as appropriate.
- Telemetry data on the other hand certainly does require access to the trusted in-car network and the traffic profiles (pseudo-real-time and short occasional or perhaps periodic messages) are arguably better suited to radio networks other than 3G.

With this arrangement the additional cost burden associated with 3G and multimedia stays with the smart-phone. This is logical in the sense that 3G (and indeed 4G) networks are designed for multimedia not telemetry-style data where the potentially vast number of connections, the short messages and associated signalling would be problematic.

With Wi-Fi network coverage as it is today there is clearly an opportunity for a dedicated low-cost and efficient technology like Weightless to handle the key telemetry-based connected car use-cases. The key will be to keep hardware costs to a minimum, and perhaps more importantly service costs so low that consumers do not care about them (or even know of them).

Now before we leave the 'who pays?' question there is one more dimension which we need to look at which relates to data in the Cloud. There is also cost associated with storing and processing this data.

A baseline model could see the car OEM owning and controlling the overall solution, including the associated costs. Just as they are responsible for the operation of the car itself, they might well take responsibility for the telemetry data associated with each car.

The initial benefit of this kind of end-to-end ownership is that it allows a very simple and controlled connected car roll-out by the OEM. Not only will the OEM be able to provide features which the fleet management industry uses today with great effect to all drivers, they can go further. They can initiate creative new business propositions with associated revenue streams using the telemetry databases. These might include:

- remote diagnostics for preventative maintenance;
- lower-cost warranty programs;
- lower-cost insurance bundled at point-of-sale;
- pay-as-you-go horsepower – rent it for a weekend trip or limit it for newly qualified drivers;
- pay-as-you-go insurance;
- green driver rebate schemes.

The potential to up-sell the data itself exists as does the opportunity to generate completely new services by making some of the telemetry information available more openly.

Having discussed trust and cost there is one further dimension which affects both and requires attention. From a cost point of view we have only considered hardware costs associated with the wireless modem but

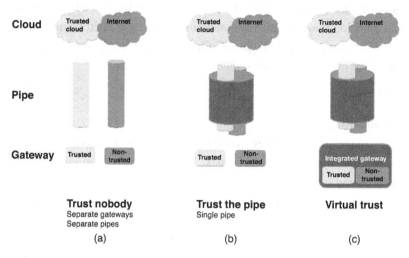

Figure 10.7 Consolidation vs. Trust

there is more to it. In Figure 10.7 we provide a more focused view of the system architecture from car gateway to cloud.

There are two stories to be told in Figure 10.7. The first relates to the consolidation of network connections (pipes), the second being the consolidation of gateway functionality.

Combine pipes?

The use-case table (Table 10.2) shows different applications and their probable trust level requirements. Physical separation is one way to ensure trust but still cater for all use-cases; i.e. different networks in the car and different wireless connections for those networks. Additional cost is inevitable with this approach. If the second connection for multi-media/infotainment is provided courtesy of a driver/passenger smart-phone then this approach could be acceptable or even sensible. We have argued that it is in fact preferred.

Combine gateways?

As mentioned above, physical separation helps with trust, so it is easy to imagine OEMs preferring two separate gateways – one connected to the non-secure Internet, the other connecting the secure vehicle network to the trusted Cloud. While it is technically possible to safely combine and partition the two gateway functions on a single microcontroller or microprocessor SoC, the argument for using the driver/passenger smart-phone for multimedia/infotainment could again prevail.

Summary

So, we have taken a journey exploring the different connected car applications and their respective virtues. We have touched on commercial viability, system architectural and trust as key dimensions. We can conclude that telemetry from the inner heart of the car can offer powerful opportunities for improving efficiency and cost for the driver and for the automotive eco-system. There are prospect of new services created on top of the car's telemetry data and resulting benefits to society. This is reflected in Figure 10.8 which shows the overall impact which telemetry can make, touching nearly all areas of the 'map'.

In contrast safety use-cases based on DSRC will be challenged due to the uniqueness of the technology and the cost of deployment on cars and on the infrastructure side. All is not lost however: radar- and vision-based solutions will come through to address a subset of the safety use-cases. Furthermore, LTE could well help with network latency.

Multimedia is best served by the 3G and LTE. For the connected car this means that the driver/passenger smart-phone or tablet, perhaps docked and connected to in-car displays, is the best fit. For trust reasons also, the total separation of non-trusted Internet-based multimedia from the in-car network is as secure as we could hope for.

OEM software use-cases are fairly straightforward in that a fixed connection (home broadband, Wi-Fi hotspot etc) will suffice and improving on an annual or bi-annual software update is easy to achieve.

What connection qualities do the telemetry use-cases require? Is Weightless suitable?

Connected car telemetry will be similar to the generally accepted telemetry characteristics; however the car will certainly be more active than, say, a domestic utility meter. Key characteristics include:

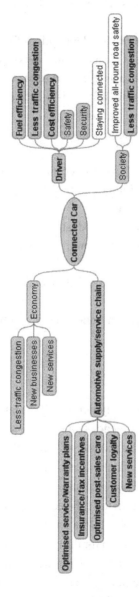

Figure 10.8 Connected car: the total value of telemetry

- Low-medium network bandwidth.
- Messages will still be fairly short but could potentially be quite frequent, as an example a message every 2 or 3 minutes would result in 20–30 messages per hour.
- Message acknowledgements will be required for some transactions.
- Pseudo-real-time will be sufficient, a few messages could be buffered in the connected car gateway for a short time.
- The connection should be trusted. This means secure authenticated access and secure network transmission.
- The car should be addressable using standard Internet protocols (IPv4/IPv6).
- Additional connected car hardware costs should be sufficiently low that they can absorbed without undue concern, by the OEM or the car buyer at the point of purchase.
- In keeping with the automotive industry, components will have to survive harsh environmental conditions and be defect free for 15 years or more.
- Network service charging should follow the eReader model – so small that you don't care.
- Coverage should be 'universal' and extend beyond national, even continental boundaries
  - international roaming
  - ideally global harmonisation.

These requirements fall centrally within the design goals of Weightless. Message bandwidth and frequency can readily support the requirements of a car, albeit the detailed data rate requirements and volumes are not clear yet. Acknowledgements are provided within Weightless. The fact that some messages can be buffered and very low latency is not required is helpful. Weightless will provide full security and IP-based addressing. Perhaps most importantly, Weightless has been designed to achieve the cost points mentioned above both for the hardware and the service charge. Finally, with its globally harmonised spectrum, Weightless is one of the few technologies able to achieve global coverage and roaming.

In summary, Weightless looks like a good solution to the telemetry requirements for the automotive industry.

10.3 Healthcare

Dr Antony Rix

The medical market is one that could benefit greatly from Weightless. There is an opportunity for technology to reduce the cost of providing care, and recent years have seen tremendous innovation in the use of wireless communication in healthcare. At the same time, it must be noted that healthcare is a highly regulated market and one that is only just beginning to realise the potential of connectivity.

10.3.1 The healthcare productivity challenge

The healthcare and life sciences market has become a major focus of attention for the communications industry during the last two decades. Healthcare in general is one of the largest markets globally, with figures indicating that spend on health is high and rising. Healthcare providers such as the UK's National Health Service (NHS) are also often amongst the largest employers. In 2009, healthcare spending as a proportion of GDP ranged from the OECD average of 9.7%, to 9.8% for the UK, 11.6% for Germany, with the highest level being 17.4% in the US. For comparison, in 1999, the average spending over the same set of countries was 7.8% of GDP [12].

This level of expenditure is clearly a major concern, with governments seeking to avoid their economies being weighed down by escalating health costs driven by rising demand, limited supply of trained personnel, increasing life expectancy and rising incidence of chronic disease necessitating more intensive and expensive care. It is also clear that healthcare is a tremendously diverse and politicised market with many different sets of needs and types of provision, which will tend to make change slow and difficult, while some have argued that major structural reforms are necessary [13].

During the last two decades, many have seized the opportunity to use technology, including health informatics and communications, to make health and social care more productive and thus reduce costs and improve the quality of care [14]. Technology corporations such as Intel,

Honeywell, Philips, Qualcomm and Robert Bosch have invested substantially in ventures targeting this market opportunity, while mobile operators and their trade grouping the GSM Association have created mHealth alliances and initiatives. Healthcare professionals are also closely involved: the largest professional body, the American Telemedicine Association, has grown to over 2200 individual members since its foundation in 1993 [15]. Many major pharmaceutical, medical device and other life sciences organisations, and a substantial number of start-ups, are actively involved in this trend.

10.3.2 The market today

Several terms are in use to describe the application of communications technology to healthcare. Perhaps the easiest to understand is telemedicine, in general the practice of medicine at a distance. Other terms have different implications, and in some cases are used to describe services that are focused on the consumer, social care or potentially disruptive models of health provision: telehealth, telecare, telehealthcare, mobile health, home health, mHealth, eHealth, connected health, wireless health.

Also worth consideration is the fact that healthcare, like many enterprises, can benefit from information technology. Conventional messaging, including email on smart-phones, and IT-based systems such as electronic health records, e-prescribing and health data management (health informatics) are now widely adopted in many markets. The smartphone offers the potential for increased usability of software applications, and healthcare-related mobile apps are well established. At the time of writing, these technologies can be considered relatively mature and widely used, in particular in the US [16], leading the Food and Drug Administration (FDA) to propose a new regime to regulate the market [17].

The focus of the remainder of this section is specifically the application of wireless communication to medical devices. An example we will use to illustrate the concepts is Cardionet which develops and markets in the

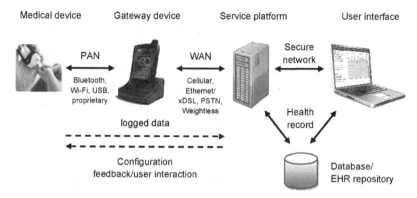

Figure 10.9 Medical devices communications via a gateway

US a wearable cardiac monitoring device that communicates electrocardiogram (ECG) data to clinical specialists using a commercial cellular network. Founded in 1999, Cardionet was floated on the NASDAQ stock exchange in March 2008 [18].

10.3.3 System architectures

Telemedicine technologies currently connect in many different ways. In the most basic case, a reading is taken on a standalone medical device then manually entered into a communication or IT system or relayed to a clinician by phone or email. This may seem to be a trivial case, but it requires no change to existing devices and remains popular. One key benefit of this simple approach, and perhaps a reason for it still being preferred by many health professionals, is that a nurse, paramedic or other practitioner takes the reading, reducing the potential for patients to misuse the device and improving the chance that other relevant symptoms may be detected.

An architecture where the human relay is replaced by a communications gateway is becoming increasingly common for devices designed to be operated by patients without supervision, as illustrated in Figure 10.9. The medical device is linked over a personal area network (PAN) to a nearby gateway device that relays sensor or usage data to a healthcare provider over a wide area network (WAN), where the data may in turn be stored in some kind of electronic health record (EHR) or other repository. This architecture is used by Cardionet and many others. Compared to manual data entry, this may improve usability, can transfer larger amounts of data like ECG recordings, and reduces the likelihood of transcription errors. There is also no reason why this approach cannot be supervised by a practitioner either at the point of care or reviewing the data in a call centre, a model preferred by the UK NHS [19].

The PAN connection between the sensor and gateway was until recently generally made by a wired cable, but now often uses a proprietary or standard short-range wireless link such as Bluetooth Health Device Profile. The WAN link between the gateway and health provider, which would previously have used the public switched telephone network (PSTN) or other wired network such as Ethernet, is now increasingly switching to wireless networks because of the ease and speed of provisioning and the desire for mobility.

An architecture similar to Figure 10.9 is described by the Continua Health Alliance [20], a trade association which promotes interoperable healthcare communications technologies based on standard data formats such as IEEE 11073 [21]. Provided it acts as a basic relay or display, the telemetry aspect of such a system is normally subject to relatively lightweight regulation both in the US and Europe – though it should be noted that the gateway, or in some cases its software, is often considered to be a medical device as the integrity of the data it passes may be critical to the treatment, and personal medical data is generally considered to be highly sensitive under data protection legislation.

The gateway architecture works well if the medical device needs to be small or battery-powered, or if there are several medical devices or other peripherals that would be used together. Indeed, if the PAN uses traditional Bluetooth or Bluetooth Low Energy, the gateway can potentially be an application running on a smart-phone, which is highly attractive to certain user groups. In some cases, the separation is blurred by directly connecting the medical device to a smart-phone through an accessory port, which may be easier to use as it can avoid the setup associated with configuring a Bluetooth network. An Internet-connected PC can also be substituted for the smart-phone, and some medical devices recently introduced to the market include a USB connector for this purpose.

Would it not be simpler to take the WAN connection directly into the medical device? This would eliminate the need for the gateway. Relatively few products for patient use have done this to date. Despite the reductions in cost noted in the introduction to this chapter, GSM and 3G technologies remain expensive when compared to the cost of goods for many medical devices like inhalers, and require substantial size and power. Ethernet or PSTN connections tend only to be directly incorporated into larger bench-top instruments designed for point of care or lab use, due to similar size and power requirements to cellular devices as well as additional safety issues. With its potential for lower size, cost and power consumption, and the benefit of providing a wireless WAN, Weightless offers an exciting opportunity to simplify the architecture of connecting medical devices, improve the usability and reduce cost.

10.3.4 Healthcare examples

What could Weightless-enabled medical devices do? The basic premise is that the device's usage or readings are logged and made available to clinical practitioners and other carers, allowing them to monitor the patient's progress and intervene early should problems occur.

Some of the simplest measurements can be useful tools for telemedicine, for example providing remote monitoring of temperature, weight, heart rate or blood pressure. Tracking a patient's weight over time can help with controlling obesity as well as monitoring other diseases and general wellness. The product here could be a Weightless-equipped weigh scales that might confirm – perhaps by audio – that the user is actually the patient, and then transmit the weight to the health provider. The combination of ensuring that the patient regularly and correctly uses the device, with personal intervention to provide the encouragement that is often needed, is key to many treatments.

An important category is more specialised devices used to monitor and treat patients with chronic (long-term) conditions. The Cardionet device noted above is primarily an aid to diagnosis of certain cardiac conditions, but there are many other types of chronic disease. In several cases adding communications to a single device offers the potential to improve the control of the condition, and some examples are presented here.

Asthma and certain other respiratory conditions are frequently treated using inhalers. A wireless-enabled inhaler can report when it is used and may also be able to sense whether the usage is successful – which depends on how the device is used. A 'smart' inhaler may also act as a spirometer, a medical device that measures the breathing condition of the patient, or a separate wireless spirometer could be provided. Data on usage can help clinicians identify if a patient is following the medication programme correctly, so they can provide training where needed. Breathing condition data or incorrect usage may also provide cues that can help identify exacerbation of the condition before it becomes severe, allowing hospitalisation or other intensive treatment to be avoided in some cases.

Type 1 diabetes is generally treated by insulin through injection or an insulin pump, with the amount and timing being adjusted depending on the patient's food intake and blood glucose level. Failure to control blood glucose can contribute to potentially severe short-term or long-term complications, but as with asthma, patients may not always adhere to the treatment very well. Providing a blood glucose meter fitted with a wireless connection can help a clinician to advise on treatment, for example by phoning the patient if a measurement is missed or is out of the safe range.

There are many other areas where communications may be of benefit. Examples include oral medication adherence, again a common cause of unnecessary complications leading to costly hospital admission, and monitoring of patients' condition at home or point of care using a range of vital signs measurements such as weight, heart rate, blood pressure and spirometry. The latter approach is being widely rolled out by a number of US health systems such as the Veterans Administration, and has recently been subjected to a large randomised trial as part of the UK NHS Whole System Demonstrators programme.

It has not been possible in this brief overview to explore the potential complications with introducing communications to medical devices. But it must be understood that embedding wireless into medical devices has implications in technical, regulatory and business aspects. It was noted above that healthcare regulators in many cases consider communications to be part of the function of a medical device, and thus subject to similar and potentially onerous rules. The cost, size and power requirements of wireless technology can also be problem. For example, generic asthma inhalers are simple, very low-cost, purely mechanical devices, and are also rather small. Adding Weightless or Bluetooth to such devices will substantially increase the size and cost, so it may be that wireless-enabled devices are selected primarily for patients who are not responding well to the established treatments.

The cost of the device and its communication function is only one part of the cost of use, which also includes providing or installing the technology, training the patient, and the time that clinicians must spend reviewing data and dealing with the results, which might include false alarms, and other system costs. While most studies of telehealth technology appear to be concluding that the technology can improve the clinical outcome, it is not so clear that it will reduce healthcare costs, as illustrated by recent studies on home blood pressure monitoring [22, 23]. One of the ways that Weightless can help with the economic argument is that its lower costs and greater ease of use should reduce the overall cost.

10.4 Should Europe's Smart Grid gain some weight by using Weightless?

Daniel Lauk, Landis & Gyr[10]

Imagine a world where enough roads are built to avoid even a single hour of road traffic jam. Imagine a world where you only get one single phone bill a year – and the amount will be based on an estimate of your last year's behaviour. Imagine a world, where airline companies would only

---

[10] Daniel Lauk is originally from Freiburg, Germany, and has been living in Switzerland since 1991. He holds a Master's degree in Electrical Engineering. Daniel started his career at Philips Semiconductors in Zurich, Switzerland. Daniel joined Landis & Gyr in 2008 as 'Head R&D E-Residential EMEA', leading communications and platform predevelopment activities for Landis & Gyr's EMEA division for residential electrical meters.

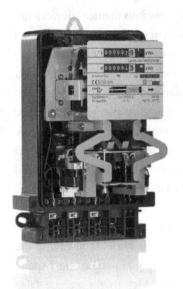

Figure 10.10 Landis+Gyr MM2000 Ferraris meter

be allowed to set one or two different tariffs – and those tariffs must be directly linked to how many miles you travel on each flight.

Sounds unimaginable, but in fact those are exactly the conditions which today's regulated electricity grids work with. In 2011, we still see many European households equipped with so-called 'Ferraris meters' (see Figure 10.10), an electro-mechanical technology that was first deployed at the end of the nineteenth century. While you are about to share a video on Facebook wirelessly via your iPhone in the living room, just one floor below you have nineteenth-century technology and market conditions controlling your electricity bill.

Deployment of smarter meters is an essential part of a Smarter Electricity Grid. Without including some sort of visibility on the last mile[11], flexible tariffing and other measures for more direct consumer interaction, the grid will largely remain what it has been for more than 100 years – a

---

[11] Typically referred to as the blind spot between a local (often street-level) transformer station and the end consumer (a household or business).

one-way infrastructure with enough generation capacity to cover maximum load plus a reserve margin. It might seem to be obvious to the reader that there is objectively a strong requirement to upgrade this infrastructure; nevertheless, due to many different reasons, market players have struggled to find the business case to do so. These conditions however are expected to change significantly and sustainably during the next decade due to the following facts and trends:

- The EU has set a mandate in their Third Energy Package[12] to install smart meters in at least 80% of European households by 2020 as a basis to build a Smarter Grid.
- The rapid expansion of renewable (hence much more volatile) energy sources, which increases the demand for maintaining excessive and expensive capacity reserve or better managing the demand side.
- The increased consumption by electric vehicles (EV's), often largely aggregated in space and time (for instance a high EV penetration in upper middle class neighbourhoods where everyone is coming home from work at the same time to charge the car).
- Greater energy awareness by consumers who request better visibility of actual consumption patterns. Transparency and timely feedback is a necessary requirement for better awareness.
- Due to an extremely high threat potential in a future smart grid, increasing security and privacy requirements are key for a successful deployment.

In conclusion, it will be vital to have an electricity metering device that is able to safely communicate consumption data in a timely manner, map much more flexible tariff schemes, allow for field software upgradeability, etc. Since only a very small portion of European households today is equipped with such instruments, there will be a need for more than 200 million additional communicating devices within the next 10+ years in Europe alone. Without doubt this will be one of the largest machine communication markets for the decade to come and it will set the pace for lowest communication costs and best possible reliability, security and lifetime.

---

[12] A set of European Union papers consisting of two Directives and three Regulations.

10.4.1 Information flow and partitioning of smart grid communications solutions

There are several types of wide area network (WAN) and neighbour area network (NAN) information flows in smart metering systems today, each of which has its own characteristics in terms of bandwidth, latency, availability and reliability requirements.

- The most frequent and biggest load is caused by billing and profile information. Typically, for such information a bulk transmission once per day is sufficient. It should be noted however that the amount of information is not constant: it will grow over time when for instance generation from photo-voltaic (PV) installations, gas consumption figures, reactive power consumption or extensive power quality data are added to the profiles.
- Smart meters must regularly be updated with configuration data such as access profiles, security profiles, tariff and pricing information. Typically such information will be updated somewhere between a few times a year up to several times per week – depending on how dynamically pricing and tariff structures are set up by the energy retailer.
- Firmware upgrades and other maintenance information will need to be transferred. While this typically is not time and latency critical, the bandwidth requirements are significant – especially in an environment where broadcasting or multicasting information is prohibited for any reason. With increasing complexity of the devices (e.g. due to security requirements), the software content will likely grow significantly, which in return will demand higher bandwidth from a communications channel for regular maintenance upgrades.
- Spontaneous device access is typically limited to few accesses per year and manual maintenance events (for instance an end consumer calling customer support where the call centre needs to get immediate access to smart meter status and billing information).
- Critical demand/response type of events. These are the only time-critical events, e.g. to shed load or trigger consumption into local storage containers to protect grid stability in a future smart grid.

Regarding partitioning of the different devices and functional units, we will see many different flavours of installation types across Europe. Most of the implementation choices will be driven by national regulation and the existing installation environment, which largely differs from country to country. In some countries, regulation is set out in a way that mandates having the communications system in a separate device (e.g. the UK), while in other countries cost targets practically mandate the full integration of communications into the smart meter (e.g. Italy, Spain, France). Due to the diversity of those installations there likely will not be globally unified interfaces between meters and communication devices – or if there are, these will be at best harmonised in a national or regional context.

It also should be noted that in most countries the energy distribution and billing infrastructure business is a regulated monopoly – a mandated or utility-driven rollout of smart meters will require the power consumed by smart grid metering and communications infrastructure to be paid by the infrastructure provider. Any cost will need to be quantified, justified and accepted by national regulatory authorities like the 'Bundesnetzagentur' in Germany – and then will be allocated to the consumer prices of energy. Similarly, in most countries regulatory authorities will control the maximum power that can be consumed by such equipment, including smart grid communications equipment. Such a regulatory environment clearly prohibits the use of consumer-type installations (like set-top boxes, reuse of Wi-Fi routers, etc.). Communications equipment also will need to sit very close to the electricity meter in order to consume energy from the unmeasured side of the household or commercial building installation.

10.4.2 Smart metering standardisation and regulation landscape

Metering and energy distribution infrastructure in general has been a strongly regulated field for more than 100 years. Due to the nature of the installations (lifetime, availability, etc.) interoperability, data privacy and robustness against tampering has been a strong request and has been a given for electro-mechanical and non-smart electronic devices for many decades – the specification of what needs to be displayed on the meter is sufficient for such devices to be interoperable with a human reader, simple sealing is enough to make sure the accuracy of the measurement is not compromised, and privacy is protected by the physical installation environment. Therefore, any future smart grid communications technology based on information technology needs to support the provision of interoperability, burglar-immunity, security and privacy. During the last year, IETF protocols based on TCP/IP have been marketed worldwide to provide a good basis for those requirements, nevertheless it shall be noted here that TCP/IP might be a good (not even necessary) *basis* for those aspects, but certainly is far from being a *sufficient* criteria.

Smart metering standardisation in Europe today is dominated by IEC standards. National and in some cases even local regulation bodies strongly influence and mandate all aspects of metrology[13], tariffing and billing. On the metrology side, there is a common European specification called Measurement Instrument Directive (MID). Any meter certified under MID in any of the European member states can be sold in all other member states without additional approvals. Nevertheless, MID only covers basic aspects of a modern meter (such as active energy for electricity metering), while national standardisations still covers wider aspects (such as reactive energy or tariffing in electricity meters). On the transport and applications protocol side the most widely used is Device Language Message Specification (DLMS)/Cosem, a universal object oriented data model language. DLMS is an IEC norm standardised under IEC 61334–4–41, and IEC 62056–53. It is not limited to electricity metering, but also covers gas, heat/cold and water metering. Corresponding standards have been developed by CEN TC294 in close collaboration with IEC TC13.

Despite the importance of unified standards for any type of long-lifetime infrastructure installation, the communications space has been widely left to manufacturers and industry associations – for instance the

---

[13] Metrology in general is the 'the science of measurement' – in the case of billing electricity meters the accurate measurement of voltages and currents according to regulatory norms. Sometimes it is expanded to the calculation of derived parameters such as total power, power factor, harmonic distortions, etc.

Interoperable Device Interface Specification (IDIS) association defining and certifying a standard for full application level interoperable and interchangeable smart electricity meters using either power line communications (PLC) or GPRS/UMTS communications.

10.4.3 Smart metering communications landscape in Europe

Today's communications technologies landscape in Europe's smart meter installations is extremely diverse. Besides the Wide or Neighbourhood Area Network communication (WAN/NAN), we also see requirements for communication between different meters (often referred to as Local Meter Network or Multi Energy Network) and towards the consumer (Home Area Network). Due to the nature of the Weightless technology and for simplicity reasons we will consider Weightless as applicable to a WAN/NAN communications use-case only.

So to summarise the principal possibilities available in today's machine WAN communications landscape and regulatory environment:

- Build and maintain a private WAN/NAN communications network (mostly Power Line Communication or RF Mesh solutions).
- Use a public WAN network infrastructure (mostly Optical Fibre/Ethernet, GPRS, UMTS).
- Use the consumer's network infrastructure (mostly DSL via in house WLAN or Zigbee connection).

While there have been some pilot attempts to use the consumer's infrastructure, future requirements will make this very difficult as it will not be possible to guarantee requirements regarding security, coverage and availability and it will be difficult to regulate – typical problems arise when the consumer has to pay for the energy bill caused by communications or regarding liabilities when consumer equipment is simply removed or unplugged. For this reason such type of infrastructure will not be further discussed here.

If we further detail private and public WAN communications networks applicable for smart metering, here is a non-comprehensive list of typical options in use today:

- Narrowband PLC in the CENELEC A band using either simple S-FSK[14] type or more complex OFDM modulation schemes (IDIS[15], PRIME[16], G3[17] to name a few). Largest scale installations up to 35 million endpoints exist today.
- Broadband PLC. Only pilot installations until date. No large scale rollouts exist or are planned in the future.
- Various RF mesh or RF point-to-point systems in different unlicensed frequency bands (169 MHz, 868 MHz, 2.4 GHz), either standard based or proprietary.
- GPRS, UMTS, CDMA or other public wireless infrastructure.

Each of today's available technologies has its own challenges and shortcomings against the most popular smart metering and smart grid use cases and requirements. Smart grid installations typically have the following common requirements:

- Simple, fast and seamless installation to keep costs low even in difficult installation environments such as basement installation or installations in metal cabinets.
- Highest possible coverage to avoid complex and expensive 'fill-in' solutions for endpoints where a chosen technology doesn't work reliably.
- Highest possible reliability – lifetime expectations are between 15 and 20 years in the field without any regular maintenance.

Within these boundary conditions, the lowest possible cost must be delivered as any smart grid infrastructure will not be perceived as added value by a consumer and must be justified and approved by national regulatory bodies. The critical factor however is not only the device cost alone, but the total cost of ownership over its lifetime; which includes cost of devices and installation, network installation and operations cost,

---

[14] S-FSK stands for Spread Frequency Shift Keying standardised under IEC 61334.
[15] IDIS stands for 'Interoperable Device Interface Specifications' and is defined and controlled by the IDIS industry association. It targets full interoperability of smart meter installations and has been based on existing IEC standards.
[16] PRIME stands for an OFDM-based power line communications standard developed by a consortium lead by Spanish Giant IBERDROLA.
[17] G3 is an OFDM based power line standard developed by a consortium lead by French Giant EDF.

cost of energy consumed by all devices along the communication path, network and maintenance cost, long-term quality or non-performance costs, etc.

If we were to come up with requirements for an ideal communications technology for smart meter installations they would have the following characteristics:

- >99.9% coverage without any manual tuning and without variations over seasons, time of day and lifetime.
- Operations expenditure for communications network significantly below 1€ per endpoint per year.
- 20+ year's availability of the network infrastructure, guaranteeing full interoperability without any field maintenance. If changes are made during lifetime, seamless firmware upgrades without changing legacy installations hardware.
- >100 kbps guaranteed raw data bandwidth with some potential to in-field upgrade capacity for future applications.
- Guaranteed Quality of Service for given latency requirements (not necessarily extremely short).
- Scalability and manageability of a large infrastructure to cover *every* household, covering everything from rural areas to dense urban areas.
- Meet state-of-the-art security and privacy requirements to protect data against abuse and protect the smart grid from malicious attacks – maybe *the* biggest challenge in a large-scale installation that is set to remain in the field for 15+ years.
- Seamless installation capabilities of communications equipment. Most European countries are covered by underground power lines today, therefore installing base stations, wireless mesh relays and alike on a large scale will pose significant contractual and cost challenges to an infrastructure provider.
- Use of a widely accepted and invisible technology – for instance in some countries significant opposition against wireless technologies can be observed once antennas become visible to consumers.

Any of the technologies used and available today significantly fails in at least one of the areas. To name a few:

- GPRS will not be available in many countries beyond 2020 and faces significant challenges in coverage, scalability and total cost of ownership. Future public wireless technologies like LTE might overcome some of the scalability and coverage issues, but remain problematic on the cost side, including the cost for endpoints.
- Broadband PLC has not delivered on communications reliability on the low voltage network and is not planned to be used in any large-scale installation.
- RF Mesh systems in today's European allocation environment faces coverage challenges in rural environments and challenges to find installation spots for mesh concentrators in urban environments. Alternatively, small cell sizes of only 10–30 endpoints lead to challenges on the cost side and require good coverage of a backbone network for master devices, such as UMTS.
- Narrowband PLC, while still considered to be the best fit today for many European environments, often has challenges in achieving the required bandwidth and future latency requirements.

Europe has already seen several full-scale smart metering installations; for instance half of the world's smart meters are installed in Italy based on narrowband PLC in the CENELEC A band. We will continue to see large-scale deployments starting over the next few years – nevertheless, the challenges in finding the perfect fit in terms of communications technology remains.

10.4.4 Requirements for a Weightless profile for smart metering

The situation described above opens up the space for new technologies as long as those new technologies can easily be applied to the commonly used protocols and security standards. Specifically, this will be the case for a regulatory environment where significantly wide new wireless bands are becoming available, e.g. due to technology migrations of other applications. One of those opportunities clearly is 'The Digital Dividend' or TV white space. It opens up a unique opportunity: in the past decades such large bands have never been made publicly available without auctioning and charging spectrum license costs (that then will need to be depreciated over network operational cost). If a nation wide coverage network can be established that realises significantly lower installation and operational costs than today's public networks and matches the coverage, reliability and cost structure of narrowband PLC systems, it is set to be a favourite for non-bandwidth critical machine (and therefore smart grid) communications.

In order to compete successfully against exiting carriers and standards (either wireless, wired or power line carrier based), a future profile for smart metering needs to have the following characteristics:

- It must be able to guarantee a minimum quality of service on both bandwidth and response time. A reasonable measure would be 100 kbps raw physical data rate and a response time <2s which is just about acceptable for sporadic and spontaneous field data accesses.
- The profile shall be able to support standard WAN transport, security and application protocols popular in next decade's smart metering world, such as TCP/IP[18], TLS[19], SML/DLMS[20] and Cosem[21].
- It shall support sporadic broadcast and multicast type information transport for applications like tariff upgrades and firmware downloads.
- Optionally, it shall support a method for very infrequent (few times/year), but response-time guaranteed reactions to singular events. This would be applicable for future time-critical demand/response applications to protect grid stability. Sub-1 s round trip latencies would be mandatory.

---

[18] TCP/IP stands for 'Transmission Control Protocol/Internet Protocol'. With the success of the internet, TCP/IP has become the universal networking protocol for computers, mobiles and other smart devices and has started to penetrate lower cost embedded devices as well.

[19] Transport Layer Security (formerly known as Secure Socket Layer, SSL) is an OSI Layer 6 security protocol widely used on the Internet, e.g. for safe payment transactions.

[20] SML and DLMS are media independent object oriented model languages targeted to smart metering applications. DLMS has been standardised in IEC 62056 and also is an integral part of the IDIS standard.

[21] The Cosem standard defines the Transport and Application Layers of the DLMS protocol. They have been standardised under various sub-standards of IEC 62056.

10.4.5 Outlook and future applications

Once smart meters are installed in every household, they will provide a unique infrastructure in a country: they offer a secure, privacy protected, reliable and low-cost communications infrastructure to every single home in each European country. To a large extent this infrastructure is financed by the necessity to build a smart grid as described in previous sections. If defined with extendibility in mind, this provides a great opportunity for other applications to make use of an existing infrastructure without having to pay for its initial deployment. This infrastructure therefore has the potential to be reused for many other new applications with critical security, privacy and reliability requirements without requiring any massive further investments. At the same time such a scenario provides the possibility for smart meter infrastructure operators to depreciate their investment with income from other markets, which again will lower the cost of smart grid operations. Examples include e-health, safety and home security, home automation and remote control, assisted living and supervision, etc. Some of those are described in other sections of this book.

10.5 Consumer applications

Dr Antony Rix

The previous sections have outlined a number of architectures, and the consumer market can be addressed in similar ways depending on the opportunity. The success of Weightless in the consumer market will hinge on how well it meets these requirements. For gateway-based systems, Weightless may be a lower-cost solution than GSM, 3G or LTE, could offer better coverage, and avoids the dependence on third-party service provision that an Ethernet or Wi-Fi broadband offering would suffer. Note however that it is likely that Weightless will be relatively narrowband. Self-contained connected devices with low data rate requirements offer the chance for Weightless to be well differentiated, provided the cost point is right. This section explores several areas where wireless could offer substantial growth potential.

10.5.1 Security

GPRS has been used for some time as a backup to a PSTN connection for security applications such as burglar alarms. Several companies have shown renewed interest in the opportunity to offer an extended service, or integrate security into a home automation solution (which is described further below). There is the potential to allow wireless door entry, analogous to car key fobs, while logging alarm, entry and exit events to the web. A system could even be used to remotely unlock the property, perhaps following a phone call from a visitor.

One of many start-ups targeting this market, AlertMe [24], introduced a security system comprising key fobs, alarm button, motion and door/window sensors, Ethernet/GPRS gateway and a plug-in unit that acts as a router for the Zigbee home area network (HAN) linking the devices. The kit is used in connection with a subscription-based web service platform. However, AlertMe did not appear to gain critical mass in this market and has since changed focus.

The author has worked with several HAN solutions that at the time of writing appear to offer substantially lower cost and better in-home coverage than Zigbee at 2.4 GHz. Combining a more efficient HAN with Weightless's potential to greatly reduce the cost of the WAN connection could help bring a proposition like this to a mass market.

10.5.2 Home energy and home automation

The smart grid section above outlined the potential for Weightless to become part of a major roll-out of communications-enabled electricity distribution and generation technology. In the shorter term, there are opportunities now to help reduce energy consumption, deliver easier control of the home, and support service business models, using wireless technology. These are also likely to be gateway-based systems, with a Zigbee or proprietary HAN, which in some cases is linked via a WAN to an Internet service platform.

Some of the most recent approaches have evolved from providing an in-home display to show users their approximate electricity consumption, helping to identify high-consumption devices and drive behaviour change, with the goal of saving money and reducing CO_2 emissions. Connecting the system to a WAN can allow a service provider to add value through tariffs or bundling, and extend the proposition to provide home automation functions such as the ability to switch appliances on or off. In some cases it may be possible to provide itemised bills through disaggregation, a method of distinguishing the power consumption of different goods. While the smart home and home automation in particular have been discussed for decades, wireless could now make it easy and cheap enough to grow substantially.

A good example of a solution today is the Green Energy Options Ensemble product, targeted at the retail energy and connected home markets [25] with detailed design work being undertaken by TTP. An 868 MHz HAN links a current clamp that monitors whole-home electricity consumption with a display and network bridge that connects via Ethernet to a broadband router. Smart plugs allow individual appliances to be monitored and remotely switched on and off through the system, from a PC or smart-phone. The system also has a Zigbee option for compatibility with smart meters, potentially allowing it to connect with the smart grid when it arrives.

Offerings of course do not have to be restricted to energy. For example, the US telecommunications provider Verizon has recently launched a subscription-based Smart Home solution with a variety of packages for home monitoring, energy and control [26].

The advantage of Weightless in these applications is that it could allow the system to be made independent of the cellular, home broadband or smart metering infrastructure, with a comparable cost of goods.

10.5.3 Smart appliances

Another long-discussed concept is to network equipment such as domestic white goods, for example the smart fridge. The cost of the technology and the disruption associated with making a wired connection between appliances and the telephone network have prevented this from gaining substantial market share, but mean there is now an opportunity for Weightless.

Energy may again prove to be a key driver for this market, as the ability to control when devices such as freezers and washing machines use energy can help reduce the peak loading on the electricity grid and thus potentially provide a cost saving to justify the cost of networking.

A WAN-equipped appliance may also enable new service business models. Just as with providing a telemetry connection in a car, this is a broad opportunity for any maintenance-intensive, uptime-critical device, which could be anything from a washing machine to an industrial pump. A service provider can benefit by selling an intelligent support contract, while the user can benefit from diagnostic telemetry enabling more rapid repairs and predictive servicing, increasing reliability and potentially reducing the total cost of owning and running the appliance.

10.6 Summary

This chapter has taken a look at four of the most promising application areas for wireless machine communications – automotive, healthcare, smart meters and consumer. These are all major applications that could see billions of devices deployed around the world over the next decade. They generate both economic value and societal value.

In practice, it is very hard to predict all the different applications that will emerge, how large each will be and its timing. Weightless could be compared to a mobile phone 'apps store' – just like the phone it provides the underlying enabler for innovators to deliver applications. And just like the apps store, it is hard to predict which 'apps' will be most popular. History suggests that there will be a large number, many of which were completely unexpected but turn out to be valuable (or fun).

Considering the applications discussed here, it is clear that there is an enormous opportunity for Weightless. All of the authors in this chapter have made it clear that for the application they were writing about there was an urgent need for wireless communications which was ill-served by current solutions and for which Weightless appeared well suited. In automotive Weightless looks like an excellent fit for telematics information which has a particular requirement for costs so low that they can be 'hidden' within other charges. In healthcare Weightless could remove the need for wireless communications from devices to have to pass through a local hub for re-transmission using cellular, simplifying devices and making them much more flexible. In smart metering there are no solutions that provide a good fit to the requirements of low cost and ubiquity – Weightless has all the right characteristics to change that. And finally, in the consumer space, the simplicity of an approach that no longer requires SIM cards and complex cellular chipsets or complex connection to a local area network could open untold opportunities. The key message is low-cost, simple and ubiquitous. As explained throughout this book these have been the design goals of Weightless from its inception.

This chapter has brought us full circle. At the start of the book we set out some requirements for Weightless and then in the bulk of the book have showed how Weightless has been designed to meet these. In this final chapter we have now demonstrated how a system with the characteristics of Weightless fits well into a number of key application areas. All that remains is to complete the standard and implement the solution . . .

References

[1] Ofcom: *Communications Market Report*, 4 August 2011, stakeholders.ofcom.org.uk/binaries/research/cmr/cmr11/UK_CMR_2011_FINAL.pdf

[2] Trevor Gill, Vodafone: *RF performance of mobile terminals – a challenge for the industry*, Cambridge Wireless Radio Technology SIG, 31 March 2011, www.cambridgewireless.co.uk/Presentation/Trevor%20Gill%20-%20Radio%20SIG%20310311.pdf

[3] www.bluetooth.org

[4] www.zigbee.org

[5] Bluetooth SIG, Headset Profile, Version 1.2, 2008, www.bluetooth.org

[6] Bluetooth SIG, 2011, www.bluetooth.org/Technical/Specifications/adopted.htm

[7] Europe's Parking U-Turn: From Accommodation to Regulation, www.itdp.org/documents/European_Parking_U-Turn.pdf

[8] media.gm.com/content/ . . . /06_22_opel_Traffic_Jam_Free_Future

[9] www.etsi.org/WebSite/technologies/DSRC.aspx

[10] 'How close are we to a crash-proof car?' www.bbc.co.uk/news/uk-15820069

[11] 'Hack attacks mounted on car control systems' www.bbc.co.uk/news/ 10119492

[12] Author's calculations based on OECD Directorate for Employment, Labour and Social Affairs, *OECD Health Data 2011*, stats.oecd.org

[13] Clayton M. Christiensen, Jerome H. Grossman and Jason Hwang, *The Innovator's Prescription: A Disruptive Solution for Health Care*, McGraw-Hill, 2009

[14] Michael J. Barrett, Forrester Research, *Healthcare Unbound*, 2002

[15] www.americantelemed.org

[16] Chris Gullo, Mobile Health News, citing CompTIA: *Half of all [US] doctors to use medical apps in 2012*, 16 Nov 2011

[17] US Food and Drug Administration, *Draft Guidance: Mobile Medical Applications*, 19 July 2011

[18] www.cardionet.com

[19] Keith Naylor, NHS Connecting for Health, *The NHS Message Implementation Manual for Telehealth*, Cambridge Wireless Healthcare & Short Range Wireless SIG, 12 March 2009

[20] www.continuaalliance.org

[21] IEEE standard 11073–20601–2008, *Health informatics – Personal health device communication – Part 20601: Application profile – Optimized exchange protocol*, 2008

[22] Stergiou G. S. and Bliziotis I. A., Home blood pressure monitoring in the diagnosis and treatment of hypertension: a systematic review, *American Journal of Hypertension*, 24(2):123–34, Feb 2011

[23] Madsen L. B., Christiansen T., Kirkegaard P. and Pedersen E. B., Economic evaluation of home blood pressure telemonitoring: a randomized controlled trial, *Blood Pressure*, 20(2):117–25, April 2011

[24] www.alertme.com

[25] www.greenenergyoptions.co.uk/our-products/energy-monitors1/ domestic-connected-home/ensemble

[26] http://newscenter.verizon.com/press-releases/verizon/2011/ the-smart-home-is-here.html

11 In closing

This book has introduced the Weightless standard, setting out the need for a wide-area wireless machine communications network and explaining how the advent of white space spectrum has provided the last piece of the puzzle needed to make such a network reality. However, the mix of very particular requirements for machine communications and the need to operate under complex rules and challenging interference within white space has led to the need for a bespoke new technology. Chapters in this book, and the *standard*, set out in detail the design decisions for Weightless, the specifics of the network, the base stations, the medium access layer and the physical layer. Subsequent chapters demonstrate how Weightless could be deployed in a nationwide network and the coverage, capacity and business case that would result from such a deployment. Finally, we concluded by looking at some of the key applications for Weightless and how they could be served by the standard.

Weightless is a unique technology. It could enable smart energy, smart cities, healthcare solutions to allow people to live at home longer and much more. It can generate excellent returns for those that make and deploy the technology and those that deliver applications using it. And, by providing the enablers to address climate change and the aging population, it might just save the world.

Glossary

ABS	Anti-lock Braking System
BER	Bit Error Rate
BPD	Burst Payload Data
BPSK	Binary Phase Shift Keying
BS_ID	Base Station Identifier
CA	Contended Access
CDMA	Code Division Multiple Access
CEPT	Confederation of European Post and Telecommunication Organisations
C/I	Carrier to Interference ratio
CRC	Cyclic Redundancy Check
DBPSK	Differential Binary Phase Shift Keying
DL_FCH	Downlink Frequency Channel
DL_MAP_LEN	Downlink (resource) Map Length
DLMS	Device Language Message Specification
DPSA	Digital Preferred Service Area
DSL	Digital Subscriber Line
DSRC	Dedicated Short Range Communications
DSSS	Direct Sequence Spread Spectrum
DTV	Digital Television
EC	European Commission
ECG	Electro-Cardiogram
ECU	Engine Control Unit
EHR	Electronic Health Record
EIRP	Equivalent Isotropic Radiated Power
ETSI	European Telecommunications Standards Institute
EV	Electric Vehicle
FCC	Federal Communications Commission

FDA	Food and Drug Administration (in the US)
FDE	Frequency Domain Equalisation
FDMA	Frequency Division Multiple Access
FEC	Forward Error Correction
FFT	Fast Fourier Transform
FRAND	Fair, Reasonable and Non-Discriminatory
GDP	Gross Domestic Product
GPS	Global Positioning System
gTID	Group Terminal Identity
HAN	Home Area Network
HOP_CH_MAP	Hopping Channel Map
HSP	Head-Set Profile
IDIS	Interoperable Device Interface Specification
IEC	International Electrotechnical Commission
IEEE	Institute of Electrical and Electronic Engineers
IETF	Internet Engineering Task Force
IFFT	Inverse Fast Fourier Transform
IMEI	International Mobile Equipment Identity
IP	Internet Protocol
IPR	Intellectual Property Rights
IPSec	Internet Protocol for Security
iTID	Individual Terminal Identity
LTE	Long Term Evolution
M2M	Machine to Machine
MAC	Medium Access Control
MID	Measurement Instrument Directive
MIMO	Multiple-Input Multiple-Output antennas
NAI	Network Access Indicator
NAN	Neighbour Area Network
NESN	Next Expected Sequence Number
NHS	National Health Service
OEM	Original Equipment Manufacturer
OFDM	Orthogonal Frequency Division Multiplexing
OMC	Operations and Maintenance Centre
OOB	Out Of Band
OSK	Orthogonal Shift Keying

OTP	One Time Pad
PAN	Personal Area Network
PAPR	Peak-to-Average Power Ratio
PHY	Physical (layer)
PLC	Power Line Communications
PSTN	Public Switched Telephone Network
QAM	Quadrature Amplitude Modulation
QPSK	Quadrature Phase Shift Keying
RF	Radio Frequency
RRC	Raised Root Cosine
RS_MAP	Resource Map
RS_MAP_MOD	RS_MAP Modulation scheme
SC_FDE	Single Carrier Frequency Domain Equalisation
S-FSK	Spread Frequency Shift Keying
SIG	Special Interest Group
SIM	Subscriber Identity Module
SN	Sequence Number
SNR	Signal to Noise Ratio
SoC	System on a Chip
SPF	Switching Propensity Factor
SSL	Secure Sockets Layer
TCP/IP	Transmission Control Protocol/Internet Protocol
TDD	Time Division Duplex
TDMA	Time Division Multiple Access
TID	Terminal Identity
T_IFS	Time for Inter-Frame Separation
T_RDD	Time to Receive and Decode Data
UL_MAP_LEN	Uplink (resource) Map Length
UWB	Ultra-Wideband
UUTID	Universal Unique Terminal Identity
V2I	Vehicle to Infrastructure
V2V	Vehicle to Vehicle
VPN	Virtual Private Network
WAN	Wide Area Network
WAVE	Wireless Access in a Vehicle Environment

Index

Printed in the United States
by Baker & Taylor Publisher Services